SMALL-SCALE LIME-BURNING

A practical introduction

Small-scale Lime-burning
A Practical Introduction

MICHAEL WINGATE

with contributions from
JONATHAN SAKULA and
NEVILLE HILL

Illustrated by
TERRY McKENNA

INTERMEDIATE TECHNOLOGY PUBLICATIONS
1985

Intermediate Technology Publications,
9 King Street, London WC2E 8HW

ISBN 0 946688 01 X

Typeset by Inforum Ltd, Portsmouth
Printed in Great Britain by Photobooks Ltd, Bristol

Contents

CHAPTER 5. KILNS

CHAPTER 6. PRACTICAL HINTS FOR KILN DESIGN AND OPERATION

CHAPTER 7. HYDRATION

APPENDICES

The Authors

MICHAEL WINGATE is an architect in private practice, a partner in the firm of Purcell Miller Tritton and Partners. He became interested in lime-burning through his work in the conservation of historic buildings and in the specification of materials for high quality new buildings. NEVILLE R. HILL is an industrial minerals geologist and adviser to UN organizations on raw materials and projects for building manufacture. (TERRE, 109 High Street, Portsmouth PO1 2HJ, UK) JOHNATHAN SAKULA is a civil and structural engineer with Ove Arup and Partners. He has been involved in small scale building materials in Africa and the Pacific.

Introduction

Lime is a remarkable and versatile material. It has a long tradition of use in building and agriculture and in many other industries; it can be made from a great variety of raw materials, some of which are abundant; and its manufacture is essentially simple and can be carried out economically on a small scale. Yet the value and potential of lime is today strangely underestimated, particularly in the building industry, where it is most needed.

More than three thousand years ago, the Cretan civilization in the Mediterranean made use of lime as a masonry mortar, and by Roman times it had become one of the basic building materials. Vitruvius, the Roman author of 'Ten Books on Architecture', written at the time of the Emperor Augustus, devotes a chapter to lime, which he says is the proper mortar material for structural walls and for ornamental plaster. In China, the use of lime for mortar and plaster is of similar antiquity. Indeed both Chinese and Mediterranean civilizations are known for the extraordinary scale and durability of the construction works which they left behind them, to which the knowledge of lime-burning made a vital contribution. And lime has continued, until present times, to play an essential part as a binding material in all large masonry construction.

In the present century a curious situation has arisen. The advent of Portland cement has provided a wider choice of building materials; and the easy availability of this material, its standardized and well-known properties, and above all its binding strength has encouraged the building industry to use it more and more in preference to lime. This has often been a mistake, and has led to over-strong mortars which cause brittleness, or porous mortars which lack durability; and it is now widely recognized by specialists that Portland cement is best used in mortars in conjunction with lime, to modify and strengthen lime, rather than to replace it. Yet still, in too many cases, the supposed need for strength is given greater emphasis than the other important properties of a mortar, and lime is omitted. There is an important educational task to be performed here, and as a wider recognition of the beneficial properties of lime in masonry work develops, the use of lime in buildings can be expected not only to survive, but to grow.

The building and construction industry is central to the demand for lime production on a small scale, not only as a masonry binder but also for use in soil stabilization, for limewash and renders, and for block-making. But there are also other uses for lime, in agriculture, in sugar manufacture, for water treatment and in the other contexts described in this book — all of which can add to this demand.

In industrialized countries, bagged dry hydrated lime is available almost everywhere, and is generally manufactured in large-scale plants. But there are circumstances — in the more remote places, or for a small local demand, or to produce special materials — where small-scale lime-burning is still both appropriate and economical.

In developing countries the scope for small-scale lime production is much greater. Throughout Asia, Africa and South America the need for new buildings of all sorts is enormous, and the scarcity and high cost of conventional building materials is presenting a real obstacle to development. In particular the very high capital cost and energy content of Portland cement is making it a high-priced material, and causing rapid escalation in building costs. Yet many of the simple building tasks for which cement is used today could as easily be done by lime or lime-based materials. Suitable raw materials are available, and both the demand and the technology are entirely suitable for small-scale localized operations. The capital cost would be low, energy would be saved, and many jobs would be created. Knowledge of small-scale lime-burning survives in a small way in many parts of the developing countries, but often the techniques used are very inefficient, and could be upgraded, and the operations expanded. In many other areas where there is a great demand for building materials, lime-burning is unknown, and could beneficially be introduced.

Since its work started in the late 1960s, the Intermediate Technology Development Group has always had a strong interest in building materials, and in particular with the problem of producing binders or cementitious materials, since this problem has always been prominent among requests for information and assistance received from developing countries. The need for a simple manual or handbook on small-scale lime-burning has long been apparent, yet not easy to meet, since the written sources of information and the expert practitioners of the art are few and scattered. In undertaking to fulfill this task for ITDG, Michael Wingate has been able to draw on the support of the members of the Group's Cementitious Materials Panel, several of whom have practical field experience. The chapter on raw materials was written by Neville Hill, and the section on wood and charcoal by Jonathan Sakula. Other panel members, notably Dr Ray Smith (of the Building Research Establishment) and Alan Pollard, late of the Transport and Road Research Laboratory, have read and commented on parts of the manuscript. John Evinson White has helped with many technical problems: the chapter on fuels in particular draws heavily on his work, and includes material produced by him; and the whole project owes a great debt to the wisdom and enthusiasm of the late Dr George Bessey of the Building Research Establishment. But the bulk of the writing, and the whole of the task of editing and preparing material for Terry McKenna's illustrations, have been the work of Michael Wingate, without whose enthusiasm and dedication over a number of years the project would never have been completed.

While the resulting volume by no means includes all that could be said on the subject of small-scale lime-burning, we feel it is both practical and up-to-date, and will be of particular value to three groups of people: those who already burn lime, and wish to find ways to improve their techniques; those who are considering starting a small lime-burning industry; and those who wish to be better informed about lime-burning whether for the purpose of advising others, or as users, or for their own interest. We hope that the work put into this manual will provide support for a steady growth in small-scale lime-burning, and increasing and better informed use of this excellent and long-used material.

Robin Spence
Chairman, Cementitious Materials Panel
ITDG
August 1984

CHAPTER 1
Background

1.1 What lime-burners do

Lime-burning is an ancient craft in which a lime-burner heats his raw material in a kiln to convert it into quicklime. He begins with some form of calcium carbonate, usually a limestone, but possibly chalk, marble, coral or sea shells. Quicklime must be handled with care as it can react vigorously with water to make hydrated lime, a process known as slaking. For many purposes the latter is a suitable form of lime and it is normal practice for the lime-burner to slake the quicklime at his lime-works.

The burning and slaking of lime are the lime-burner's central activities, but to make the operation worth while he may become involved with quarrying, crushing, grinding, mechanical handling, bagging, storage, marketing and distribution of several products. All of this will need management.

The scale of these operations was once very modest but in the last hundred years the equipment used in the industrialized countries has become large and expensive. In many situations it is still appropriate to operate on a small scale.

This handbook will describe how lime-burners can make the best use of available resources in small scale operation.

1.2 What lime is

The quicklime which the lime-burner draws from his kiln is the lively material which was once familiar on building sites where the lumps would hiss and crumble in the slaking pit. Quicklime is the oxide formed by driving off carbon dioxide gas from calcium carbonate or from dolomite.

If the quicklime is pure and correctly burned the lumps will fall vigorously to a dry powder when water is added — the water combines chemically with the quicklime. The most widely available form of lime is this 'dry hydrate' which is usually just called hydrated lime.

If more water is added the hydrate will 'run' to a lime putty. This is a soft and smooth material which can be stored indefinitely if it is kept wet.

Add still more water and the putty will become a milky suspension known as 'milk of lime'. This is used in some industrial processes where the lime must flow through pipes, but it is too bulky for transport or storage.

If the suspension is allowed to settle, a clear saturated solution is left above the putty. This is lime water.

Few limestones are pure calcium carbonate. Many contain clay impurities and these have a marked effect on the lime which will be produced. If the proportion of clay is very high the lumps of lime will not readily slake and the quicklime must be finely ground before it can be used as cement. Such limes are suitable for building but not for other uses. The mortar which a builder makes from them can set under water and these limes are known as 'hydraulic limes'.

There is a full range of limes which may be usable between this extreme and pure, non-hydraulic, limes. Normal commercial lime-burning is restricted to the pure limes, but some of the other classes of lime which are not normally available are very suitable for building.

There are many terms used to describe lime in different forms and qualities and these are listed in the glossary.

1.3 Uses of lime

The first important use of lime was in the building industry but this now accounts for only a small fraction of lime produced in industrialized countries. The chemical and physical properties and the low cost make lime an essential ingredient in many industrial processes.

Lime still has an extremely important role in good building and this may provide the full market for many small lime-works, but the lime-burner needs to know what other opportunities exist. The largest industrial users of lime generally have their own or 'captive' lime-kilns and only buy small amounts on the open market to balance their production and demand. These industries are steel-making, especially with the Basic Oxygen process, sugar production from sugar beet, large-scale paper making, the chemical industries and the manufacture of calcium silicate bricks.

A recent development which is particularly effective in hot

climates is the use of lime to stabilize soil in the preparation of road foundation bases and air-fields. These projects demand quite large amounts of lime in a short delivery period and would not fit in conveniently with the operation of a permanent small-scale lime-works. But a similar use is the manufacture of lime-stabilized, compressed soil blocks. In hot climates these can be more durable than traditional soil blocks.

Although sugar beet refineries make their own lime, sugar cane producers may well buy quicklime from a small lime-works. About 30 kg of quicklime is needed to make each tonne of cane sugar.

Lime plays an important role in most of the metal extraction industries. It is used in beneficiation of ores and to neutralize acidic gases in the smelting of copper, zinc and lead. Other industrial uses include the tanning of leather, glass making and the preparation of certain bleaches, dyes, fungicides and insecticides.

Lime was once used on farms to neutralize the soil, but although farmers still talk of 'liming' the land what they actually use is crushed chalk or limestone which can achieve the same effect more economically. Crushed chalk or limestone can be a useful by-product of a lime-works. In market gardens lime is still used to adjust soil acidity and to help with the action of fertilizers. Quick-lime sometimes has to be used on farms to destroy the infected carcasses of diseased animals. Disease can sometimes be avoided by painting farm buildings with limewash — a cheap form of paint which is a mild germicide.

Lime is used for softening water and can be used prior to chlorine in water purification. It is used in both small-scale and large-scale sewage treatment and for the treatment of many industrial wastes. In some countries the balance of nature is being destroyed by 'acid rain'. Attempts are made to correct this by adding large amounts of lime to lakes and rivers.

1.4 Lime as an alternative to Portland cement

During the nineteenth century new types of cement were developed. They were stronger than most forms of lime and although rather expensive they were strenuously promoted in competition against lime. Even though the new cements had severe disadvantages for some tasks, they came to dominate the market. National standards were drawn up and the name 'Portland cement' was used to describe the carefully controlled products which could conform to these standards.

The promotion of Portland cement has been so successful that it is widely forgotten that excellent standards of building are possible using limes instead. In some parts of the world Portland cement is difficult to obtain and the possibility of using locally produced limes for building should be considered even if no lime industry exists.

Lime was always used for the mortar in brickwork and stone masonry. All forms of lime are suitable although in harsh conditions the hydraulic limes are more durable.

Lime is much better than cement in plasterwork. The setting is slow, but the result will look better and the softer surface will be less likely to crack. Lime plaster is particularly suitable for decorative work.

Portland cement is essential for highly stressed concrete, but, for work at low stress, hydraulic limes can make good concrete. This is quite suitable for many footings and ground slabs. It has been used very successfully for vaulted work.

One of the simplest forms of paint for use on walls and ceilings is limewash. This is easily made from quicklime or from lime putty. It is a good alternative to cement-based paints.

CHAPTER 2

Lime-burning: Theory and Practical Constraints

2.1 Dissociation

The essential chemistry of lime-burning could hardly be simpler. When calcium carbonate is heated it dissociates to form calcium oxide (quicklime) and carbon dioxide gas.

$CaCO_3$ + Heat $\rightleftharpoons$ CaO + CO_2

calcium carbonate $\rightleftharpoons$ calcium oxide + carbon dioxide
(limestone etc) (quicklime)

The approximate molecular weights for this reaction are

$100 (CaCO_3)$ $\rightleftharpoons$ $56 (CaO) + 44 (CO_2)$

so, in round terms, the carbonate yields about half its own weight as quicklime.

The various physical forms of carbonate begin to dissociate at different temperatures. Calcite, for example, begins at around 900°C. When a lump of carbonite is gradually heated in a kiln it dries out whilst its temperature rises to the dissociation level. The outside of the lump will be hottest and will decompose first. As the decomposition spreads inwards the carbon dioxide must diffuse through the pores of the lump. If the carbon dioxide is not removed the reaction comes to a standstill and will begin again only if the carbon dioxide pressure is reduced or if the temperature is raised. A graph showing this relationship between the equilibrium dissociation pressure and the temperature is included as Appendix 4, but it should be remembered that it is extremely difficult to assess the temperatures within the materials.

2.2 Reactivity, underburned and overburned Limes

Just as the carbonate has several structural forms, so has the oxide. The most reactive forms of lime are produced when the burning is kept to as low a temperature as possible. This is achieved without effort when wood is used as the temperatures barely exceed 900°C. For lime to be produced at these low temperatures the partial pressure of the carbon dioxide must be kept low and the steam produced by burning wood helps to achieve this.

If the burning is stopped before the decomposition has reached the centre of a lump there will be a residual core of carbonate. The lime is said to be underburned. In extreme cases the lump will feel too heavy when handled.

In normal commercial practice kiln temperatures are kept high to speed up the decomposition. Temperatures around 1250°C are quite normal. 'Solidburned' quicklime formed at this temperature has a more formal structure and is less reactive than the loosely structured 'lightburned' lime produced at low temperatures.

Above perhaps 1250°C there is a danger that slight impurities in the raw material can form fusible products which close the pores of the lumps. The lime is then said to be 'overburned' and will have a wizened appearance. The closed structure makes the overburnt quicklime very slow to slake and this can cause serious problems in some uses.

At very high temperatures the lime can become 'deadburned'. In this form the calcium oxide has almost no reactivity. Indeed, deadburned dolomite is so stable that it is used as a refractory material.

There is a relationship between kiln temperature, limestone lump size, and time needed for calcination. This was explored by C.C. Furnass and a graph based on his results is included in Appendix 3.

2.3 Gas flow through a kiln

When a lime-kiln has reached its normal operating temperature the carbon dioxide gas must be removed from the lime as it is evolved. If this were not done the reaction would come to a standstill or even reverse. An adequate flow of gases through all parts of the kiln is essential to achieve this. The gas flow serves another major purpose and that is the transfer of heat through the various zones in a kiln. As every lump must be heated to the same extent to yield its lime the gas flow must be even over the whole cross-section.

The gases flow through the network of voids left between the lumps. When perfect spheres are packed closely together these voids represent one third of the volume of the solids. In practice the lumps of lime will be graded so that the smallest lumps are not less

A kiln cross-section showing an even gas flow through an open network of voids.

than half the size of the largest and their shapes will be quite irregular. These variations in shape and size tend to reduce the proportion of voids if the material is closely packed, but this is largely balanced by the looseness of the packing.

The ratio of about 1:3 remains constant whatever the actual size of the lumps, but gas flows more easily through larger voids than smaller spaces. This is because of the friction effect at the boundaries between the solids and gas. With smaller lumps the greater surface area gives greater friction.

If the lumps were not graded carefully the smaller pieces would tend to fill the voids between the larger pieces blocking the passage available for gas flow. This must happen to some extent in kilns where coal is mixed in with limestone, but the problem can be very serious if the materials in the kiln are mechanically weak in either the carbonate or oxide states. If the materials crush or degrade the gas flow pathways can become completely blocked. A guide to the pressure drop which might be expected can be taken from Appendix 6.

The gases will tend to flow upwards through natural convection. Sometimes the flow is assisted by mechanical fans either at the inlet or exhaust of the kiln, but in many cases the natural draught can be exploited. This cannot be accurately predicted but a guide relating the draught to the mean gas temperature and effective height is given in Appendix 7. A chimney may be built on top of the kiln to increase this natural draught, and in every case a control will be needed to give sensitive adjustment of the draught. This could be a control on inlet or exhaust flow.

Shells and soft chalks are forms of carbonate which pose special

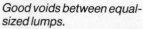

Good voids between equal-   *Small pieces filling the voids.*
sized lumps.

problems for the kiln designer. Flat shells pack so densely that they are usually burned in wide shallow pans which make heat recovery very difficult. Shaft kilns designed for soft chalks are usually limited in height to around 12m to restrict the crushing pressures.

2.4 Heat requirement and fuel economy

A mean figure for the heat required to dissociate the carbonate is shown in Appendix 8. The purer forms need rather more than the impure forms. The figures given are the unattainable theoretical targets which the lime-burner must aim at. They show the energy required to change from the carbonate to the oxide under ideal conditions.

As well as the heat essential to effect the dissociation, heat is also required for each of the following:

To raise the fuel to its combustion temperature.
To raise the temperature of the air flowing through the kiln.
To raise the temperature of the fabric of the kiln to its operating temperature and to maintain this temperature against the heat losses from the outer surface.
To drive off any water in the carbonate
To raise the temperature of the carbonate to the required level.

As the heat must flow by conduction into the centre of each lump the surface temperature must be above the dissociation temperature. If the surface temperature is only slightly higher (as in a kiln fired by wood) the penetration of heat to the centre will be relatively slow. Smaller pieces will heat through more readily than large lumps and this is further reason for grading the carbonate carefully. In Appendix 3 the relationship between time for dissociation and particle size is shown for various temperatures. The effects indicated for very fine particles are of particular interest. Conventional kilns need to use rather large lumps to achieve an easy gas flow, but recent experimental work on dissociation of fine particles at 'low' temperatures show how excellent lime might be produced with good fuel economy.

In any kiln the lime is produced at high temperature and it is generally possible to recover most of the heat stored in the freshly burned lumps. The usual technique is to cool the lumps by passing air over them and using this preheated air to burn the fuel. The exhaust gases from the combustion zone (including any unburnt air) then pass on through the carbonate giving up much of their heat to preheat the carbonate. The details of this heat recovery vary considerably from one class of kiln to another, but nearly always imply a continuous production process in which the charge flows (if at all) downwards while the gases flow upwards. These continuous processes also avoid the need to repeatedly heat up the fabric of the kiln so that only heat losses are through radiation from the outer surface, through residual heat in the exhaust gases and through residual heat in the cooled lumps. The first two of these losses are proportional to the time taken for the dissociation, and this may be a reason for the usual commercial practice of burning lime quickly by opting for the higher gas temperatures. Radiation losses will be reduced by good insulation and low surface area. Losses in the exhaust gases will be minimized by careful draught control by the lime-burner.

Although continuous production methods offer the best fuel economy, other reasons may require a batch production. In these cases heat recovery can sometimes be achieved by designing the operation as a short sequence of continuous production, as in the climbing kilns described later. For batch production the kiln designer must arrange to keep the thermal capacity of the kiln as low as possible and the insulation must be of the highest standards. Again the lime-burner must control the draught carefully and he must be able to judge when the burning has been completed.

One of the greatest areas for effective fuel economy is in the correct burning of a suitable fuel. The matter is considered in Chapter 4 but this is an area where further advice must be sought. In broad terms a long cool flame is needed. This is readily available from timber and to a lesser extent from long-flame coal. Where other fuels are used steps must be taken to achieve a good flame spread and to reduce hot spots which might cause overburning of the lime. The composition of the exhaust gases will show whether the combustion has been efficient. In a large lime-works gas analysis will be used, but on the small scale the lime-burner will develop the necessary skills to judge the gases.

CHAPTER 3

Raw Materials

Lime is made by burning limestone or other materials that are composed mainly of calcium carbonate. They exist in a wide variety of forms. These range from recently deposited oyster shells and coral, through consolidated sedimentary limestones and marble which is a metamorphic rock, to carbonatite, a rock now generally considered to have been of magmatic (igneous) origin. In addition there is calcium carbonate sludge, a by-product from water softening treatment and from the sulphite paper-pulp process, which is utilized to regenerate lime for re-use in these chemical processes. There are also naturally occurring deposites of unconsolidated calcium carbonate which can be handled in the same way as these sludges.

Although calcium is the fifth most abundant element in the Earth's crust, and practically every country has deposits of limestone, it is not always possible to find occurrences of the rock which are suitable in chemical purity, physical characteristics and location, in the particular area where it is proposed to establish the manufacture of lime.

Before dealing with how to find and test a possible source of suitable raw material, this chapter will first mention the types of limestone that may be available and then show how they are likely to perform in producing the quality of lime sought by the potential users.

3.1 Classification of limestones

The lime-maker is really only concerned with the chemical nature of the limestone and certain physical features such as its hardness and the shape of the stone when broken. The geologist, or whoever is charged with finding and establishing a suitable supply of limestone, will usually need to understand the origin or mode of formation of the rock. This is so that the search area can be concentrated where the stratigraphic (geological historic) conditions for limestone formation were most suitable and to allow an assessment of the likely quantity of material and the way that it should be quarried. A detailed account of the origin of limestones is not necessary here but is available in geological and petrological textbooks. Some of the classifications used by geologists are included here so that the lime-maker is aware of the variety of limestone types and some of the terms used to describe them.

The pure mineral form of calcium carbonate ($CaCO_3$) is calcite, of which a well-known variety is Iceland Spar. Many limestones are calcitic though the more recently deposited may consist of a somewhat denser and harder crystalline form called aragonite. Frequently, magnesium carbonate ($MgCO_3$) is also present so that the rock is either a dolomitic limestone or dolomite ($CaCO_3$, $MgCO_3$) in which the proportion, by weight, of the magnesium carbonate is over 40% and the remainder is calcium carbonate. The pure mineral dolomite contains 45.7% magnesium carbonate.

Some authorities distinguish between limestone and dolomite. Here, any rock whose main chemical constituent is calcium carbonate will be considered to be a limestone and this will include dolomite which even in its pure form contains approximately 54% $CaCO_3$.

To show the variety of rocks that the term 'limestone' embodies, the following classifications by Dixey are useful:

According to origin:

i) Organic   Marl, chalk, calcereous ooze, fossiliferous limestone (including shell limestone and coral limestone etc).

ii) Chemical   Compact limestone, marble, stalagmite and stalactite, concretionary limestone, kunkar, oolitic limestone, calcareous tufa or travertine.

According to composition:

Argillaceous or clayey limestone, arenaceous or sandy limestone, siliceous limestone, ferruginous limestone, phosphatic limestone, magnesian limestone, Dolomite, etc.

According to texture:

Compact limestone, earthy limestone, cherty limestone, conglomeratic limestone, modular limestone, crystalline limestone or marble, saccharoidal limestone, oolitic limestone, unconsolidated calcium carbonates, etc.

Another classification into three groups, used by geologists and quoted by Boynton and Gutschick, is according to origin:

1 Autochthonous (or accretionary) limestones which grew in place by chemical or biochemical processes that extracted calcium and magnesium carbonates from seawater. Chalk is an example. Calc tufa is deposited from supersaturated inland waters.

2 Allochthonous (or detrital) limestones, where the constituent material, consisting of fragments of coral reef, shells and other fossil debris, has been transported and redeposited by water currents so that the rock shows clastic rock features such as current bedding and sorting. They subsequently harden by compaction, cementation or crystallization.

3 Metasomatic limestones, where the original character of the rock has been modified by secondary impurities brought in. But the phosphate, iron, silica, etc., may also have been there originally to form the phosphatic, ferruginous, siliceous or cherty, argillaceous, asphaltic, etc. limestones.

The value of lime for most of its uses depends mainly on its content of 'available lime' (calcium oxide or calcium hydroxide). Magnesia is frequently the impurity that affects lime in manufacture and use, so the classification of limestones for chemical processes, such as lime and Portland cement* manufacture, is based on the contents of $CaCO_3$ and $MgCO_3$:

Ultra-high calcium limestone: more than 97% $CaCO_3$
High calcium limestone: more than 95% $CaCO_3$
High purity carbonate rock: more than 95% $CaCO_3 + MgCO_3$
Calcite limestone: less than 5% $MgCO_3$
Magnesian (dolomitic) limestone: 5–40% $MgCO_3$
Dolomite: more than 40% $MgCO_3$
High magnesium dolomite: more than 43% $MgCO_3$

*for Portland cement, the $MgCO_3$ has to be less than about 6–7% (3–3.5% MgO) so that the resulting level of MgO in the cement is below about 4–5% to avoid problems of 'unsoundness'.

3.2 Limestone as raw material for lime-making

A limestone will only be a suitable raw material for lime-making if it can yield the quality of lime needed by a particular consumer and if it has suitable physical properties for use in the available types of lime-kiln.

Limestones having high $CaCO_3$ contents, well over 90% and preferably as much as 98%, are sought as raw material for lime for all three main market sectors. These are industrial or chemical purposes, building and construction use including soil stabilization and in agriculture as an acid soil neutralizing agent, fertilizer and general soil conditioner.

The industrial consumers, such as the sugar refiners, paper manufacturers, etc., need lime that has a high content of available lime as calcium hydroxide. The lime-plants that intend to supply it must have ready access to high calcium limestone, preferably, or else high purity carbonate rock.

Lime for the manufacture of building products such as sand-lime (calcium silicate) bricks should also, preferably, be made from these high calcium carbonate limestones. But the ordinary building work on site, such as brick-laying, masonry, plastering and rendering can be done as well or better using 'grey limes'. These hydraulic, or semi-hydraulic limes, so called because they will harden under water, are made by burning argillaceous (clayey) hydraulic limestones which will be discussed later. Unfortunately, just as millions of people have come to believe that white bread and sugar are 'better' than brown, so white lime tends to succeed in the market over grey in the developing countries, as it has already done in the industrialized nations. Dolomitic limestone is excellent for making building lime provided that one ensures that no unslaked magnesia, as periclase (MgO), remains after hydration and before the lime is applied. Otherwise, the delayed hydration of the magnesia causes mortars to expand the plaster surfaces to develop pitting and popping. The problem can be overcome by pressure hydration in an autoclave or else by keeping the lime covered with water for many days in a slaking pit. In fact, Pliny the Elder, in AD 79, claimed that no lime should be used until it had been slaked for three years, possibly for this reason as well as to increase its plasticity.

To be most effective in its many roles in the soil, lime hydrate needs a high level of available lime as calcium hydroxide and so it also should be made from a high calcium limestone. In general its effectiveness decreases as its purity decreases though some of the impurities, particularly phosphate and any potash, may be beneficial. Although many soils for particular crops, such as tobacco, require additions of magnesia, this is normally applied as unburnt dolomitic limestone, finely ground, and sold as agricultural lime. This is another case where caution has to be taken in the use of lime made by calcining and slaking magnesian, or dolomitic, limestones, as the magnesium hydroxide is considered to be too caustic for organisms in the soil.

Thus in the selection of raw material for lime-making the requirements of the intended market have to be kept in mind. For ordinary building purposes, for instance, the less pure limestones may prove to be quite satisfactory. The same is the case for Portland cement manufacture where as little as 70% $CaCO_3$ is acceptable in the limestone provided the balance of other components is appropriate.

The purest forms of calcium carbonate, such as chalk and calc tufa, are white or nearly so. However, colour may be quite misleading as an indication of purity. Some of the whitest 'limestones' are nearly pure dolomites, which contain no more than 54% $CaCO_3$. Many highly calcitic limestones, with over 90% $CaCO_3$, are grey. The principal limestone for lime making in England is the carboniferous limestone which is grey even though it contains 98.9% $CaCO_3$. In places it is bituminous and this can make it a very dark grey.

To make high quality lime the chemical composition of the raw material must be right. To be suitable for processing, particularly in shaft kilns, the physical nature of the limestone has to be right, too. (In Portland cement production this is less important as there the limestone is milled to a fine powder before it enters the kiln, blended usually with clay or shale.) Ideally, the limestone should break into roughly equidimensional lumps when being prepared for the kiln and should remain as lumps until it leaves the kiln as quicklime. Depending on the type and size of kiln, the limestone is prepared to a particular range of size which is within fairly narrow limits. It is important in all kilns to have raw material that can establish and maintain a certain size and shape, so that the stone is evenly burned. In rotary kilns it is not too critical if the stone, which starts off at a certain size, say $10 \times 20mm$ or $20 \times 40mm$, becomes smaller during firing so long as this happens fairly uniformly and segregation is avoided. In vertical shaft kilns, however, there is an optimum stone size that has to be maintained, not only to promote uniform burning but also to permit adequate draught up through the column of limestone in the kiln. If the limestone is soft and friable or shatters into small pieces (decrepitates) during firing, then other firing systems have to be used, such as the Fluosolids kiln, or else the limestone must be milled and pelleted with a solid fuel such as coal. On a small scale, batch firing in a pit or open trench kiln may be the only way.

Vertical shaft kilns are relatively easy and inexpensive to build and have good fuel efficiency when operated continuously, so in considering the ideal type of rock and how various types of limestone will perform, it is presumed here that a vertical shaft system, either mixed feed or oil or gas fired, is to be used. The raw material for this, then, should be a fairly hard limestone that resists abrasion, has uniform porosity, is fairly massively bedded and jointed rather than thinly bedded and flaggy, and is able to withstand thermal shock without breaking up into small pieces. In addition, depending on the market, it should be a high calcium limestone or high purity carbonate rock or can be, for ordinary building purposes, an argillaceous limestone with, say, up to 25% of clay mineral, or alternatively, up to about 10% silica, SiO_2, and 5% alumina, Al_2O_3, if chemically analyzed.

The occurrence and performance of some of the raw materials used in lime-making will now be considered in relation to the foregoing discussion on what is required of a raw material.

Shells and coral

These have the advantage of being very pure forms of calcium carbonate, except when the shells have sandy encrustations. Also their location on the coast line is often conveniently close to centres of habitation and they are in an accessible and transportable form, though diving may have to be employed. The shape of oyster shells, clams and to some extent conch shells, is not suitable for vertical shafts kilns and usually they are burnt on a small scale in pit or beach kilns. Some coastal-based chemical producers in Texas and Louisiana have burned oyster and clam shells in rotary kilns. Coral is also of poor shape for shaft kilns and its variable porosity means that it burns unevenly. In some areas, such as the Eastern Caribbean, this use of coral is causing concern on environmental grounds. Governments are likely to prohibit removal of living coral there and in the Pacific islands as well.

Fossiliferous limestones

These include coral reef limestones and shelly limestones as well as finer grained types made up of foraminifera, coccoliths, etc. Unless the rock is well compacted or has become recrystallized, the coarsely fossiliferous rocks may crack along the fossil boundaries when they are fired. Firing tests may prove otherwise. As coral lives in clean, rather than muddy, waters, these reef limestones are usually fairly pure and massive and make good raw material.

Chalk

This is also usually of organic origin. It is very pure, often containing more than 98% $CaCO_3$, white and of such a fine grain size that at one time it was thought to be amorphous. It consists of the fragments of tiny marine organisms (coccoliths) with a flower-like structure with 'petals' only about 1 micron (0.001mm) in size. These are what give chalk whiting such good properties as a filler. Chalk can make an excellent lime provided it is not too soft. It is still used in some shaft kilns in southern England, for instance.

Chemically precipitated limestones

Some of these, such as calc tufa and travertine, have been deposited from springs of hot water whilst stalactites and stalagmites, which grow as 'icicles' of very pure calcium carbonate in caves, are derived from cold waters passing through limestone areas.

Travertine

This is much used by operators of small lime-kilns in the Andes of Ecuador, for example. It occurs spasmodically, usually as mantles hanging down the side of a steep valley where a now dried up warm spring disgorged itself as a waterfall. Caution is needed in estimating the reserves of limestone represented as they usually form a cap only a few metres thick but appear to be thicker. Whilst, like the stalagmites and stalactites, they are hard, the travertine of the type famous around Rome as a building stone sometimes has irregular porosity which can lead to uneven burning. Thus the denser parts may leave a core of unburnt limestone when the more porous parts are completely burned to quicklime. The Romans found the best lime came from hard and compact travertine, rather than soft and porous tufa.

Calc tufa

This is the general name for deposits of calcium carbonate which precipitate from 'hard' water, i.e. containing calcium bicarbonate, and are analogous to the deposits of 'fur' that sometimes build up in kettles and hot water tanks.

Calcrete

This is a surface layer of gravelly material that has become cemented together by calc tufa. Depending on its purity it may make lime adequate for ordinary building use, providing the coarser gravelly pieces are sieved off after slaking the lime. Then the remaining sand size material stays with the lime to make up a building mortar.

Dolomitic limestones and dolomite

These rocks were usually laid down as calcitic limestones which were changed to the double carbonate, $CaCO_3.MgCO_3$, by magnesium-rich brines. Some dolomites, though, have been deposited by evaporation. The process of dolomitization tends to obliterate fossils that were present. Lime-makers prefer burning these rocks as less heat is required to calcine dolomite crystals than crystals of calcite. The theoretical dissociation temperature of calcite limestone is 900°C, but that of dolomite is only 725°C.

Operators of large modern lime-plants have no problem in using magnesian limestones as they can ensure that no magnesium oxide (MgO) is left in the hydrated product by employing pressure (autoclave) hydration. Much of the lime in industrialized nations is used for steel making purposes where MgO is desirable. The usual use of lime in developing countries, though, is for building work. There, residual MgO in the lime is very undesirable. Pressure hydration is often not possible and the lime-maker has a slow hydrating quicklime which is not nearly so easy to use as a soft burned, reactive, high calcium lime.

As mentioned earlier, magnesia should not normally be added to the soil in the form of the oxide or hydroxide and so in agriculture the magnesium requirement of the soil is usually applied as ground magnesian limestone. This, then, is a further possible disadvantage for the lime-maker who uses dolomitic limestone or dolomite.

There are conflicting opinions as to what level of magnesia is desirable in lime for processing cane sugar. The U.S. Bureau of

Standards stipulates a maximum of 3% MgO in hydrated lime for use in the manufacture of sugar, whilst much higher levels are claimed to be practicable by at least one U.S. lime manufacturer. For the lime-maker intending to supply the local sugar industry in a developing country, it is safer to avoid the dolomitic limestones and use, if available, a high calcium limestone.

Argillaceous and other impure limestones

Other than magnesia, which is not necessarily an undesirable constitutent, the common impurities in limestones are clayey i.e. argillaceous (alumino-siliceous) material, silica as chert nodules, quartz veins or sand grains, ferruginous (iron-bearing) minerals such as haematite and pyrite, phosphate due to the presence of bones or the calcium phosphate mineral apatite or guano, carbonaceous matter such as bitumen, or as graphite in some marbles, and sulphur present as pyrite, FeS_2.

The presence of these materials obviously reduces the percentage of calcium carbonate in the limestone and the possible level of available lime in the product. The latter is also reduced by some reaction with CaO. Silica, for example, combines to form calcium silicate. So the chemical efficiency of the lime in industrial processes is much weakened, as also is its effectiveness as a soil conditioner. Sulphur is harmful in lime for steel processing where it should not reach above 0.04%. Except when limestone is used for Portland cement production, where it should be kept below about 2 to 2.5% in the raw mix, phosphate (P_2O_5) does not seem to figure as an objectionable contaminant.

Hydraulic limes are grey coloured and are often sold as 'grey lime'. They harden more strongly than do the 'fat' or white, high calcium limes and can even set under water. They can be made from calcitic limestones that contain up to about 25% silica and alumina in a finely divided and dispersed form, such as the clay minerals present in an argillaceous limestone. Vicat suggested the following correlation of degree of hydraulicity with content of clayey matter:

Active Clay Content (%)	Designation
less than 12	feebly hydraulic
12 – 18	moderately hydraulic
18 – 25	eminently hydraulic

Cement rock or stone is an argillaceous limestone having around

70 to 75% $CaCO_3$, such as the Jacksonberg Formation of the Lehigh River area of Pennsylvania. It requires little or no adjustment for use as a raw materal for Portland cement manufacture. It would make an eminently hydraulic lime if fired at the lower temperatures used in lime-burning.

Marl, which is a calcareous mudstone, may also be successful for making grey building lime, though often the $CaCO_3$ content is inadequate and the rock is too soft for successful shaft kiln operation. Unconsolidated calcareous deposits are occasionally very pure and may be wrongly called marl.

Marble

In geological terms marble is a metamorphic rock. It is a limestone that has been subjected to pressure and possibly also heat so that the rock has become wholly crystalline, hard and dense — and thus able to be cut and polished as 'dimension' stone. However, the term 'marble' is applied by the layman to many other rocks such as travertine, that are not metamorphic, but that accept a polish.

To the lime-maker the attraction of marble is that it often has a high carbonate content, many marbles being dolomitic besides calcitic. It can usually be crushed into roughly equidimensional stones and it has a uniform, though dense, porosity. Its disadvantages are that the low porosity makes it less easy to calcine and the impurities, such as olivine minerals, are too coarsely crystalline to react and make a hydraulic lime. Graphite and pyrite are other common impurities.

Several authorities advise that coarsely crystalline limestones are more likely to decrepitate and so be unsuitable for shaft kilns. Though in general this may be so, it is always worth carrying out trials in a small kiln to examine the behaviour of such stone. Trials in Malawi, for instance, indicate that the very coarsely crystalline calcite and dolomite marbles of the Basement Complex (Pre-Cambrian) retain their lump form when calcined. Upon hydration, though, they do not readily produce a hydrate of high specific surface area which gives the high plasticity that is sought in building lime. However, the problem may respond to improved methods of hydration, in place of the traditional methods being employed.

Carbonatite

Little is known about the performance of these magmatic rocks which were used as raw materials for lime-making in Malawi, for instance, and also in Uganda where they provide the raw material for cement manufacture. Sovite, the rock present in the alkali igneous rock ring structures of southern and eastern Africa, appears to have several good features. It is of high calcium carbonate content, hard and massive and, depending on location in relation to markets, well worth investigating as a raw material. Some of the deposits have big reserves.

3.3 Prospecting

Before setting out to look for a suitable deposit of limestone, a prospector has to have a good understanding of the extent and thickness which are likely to be needed. This will also help in discussion with the staff of the government's Geological Survey Department. Otherwise, there is the risk of either siting the kiln by a deposit that is exhausted after a short while or else wasting valuable time and scarce money in proving reserves far in excess of what can possibly be used by the project.

The lime-kiln, or kilns, will normally be located as close as possible to the quarry. It is usually wasteful to transport the lime as limestone because it contains up to 44% of carbon dioxide which will be lost, together with moisture, as gases out of the top of the kiln.

Several factors are taken into account in calculating the required reserves:

Calcination and hydration

These are reactions that reduce and increase, respectively, the mass of the material being processed. Pure calcium carbonate loses approximately 44% and pure calcium oxide gains 32% in weight when fully hydrated to dry calcium hydroxide powder. In practice, of course, the limestone is never pure; the gains and losses will be in proportion to the purity. It is fairly easy to predict the likely weight loss on fully calcining the rock, but the weight of hydrated lime produced is less than would be predicted because the coarse particles must be removed. These include underburned limestone as well as any unreactive quicklime. They are removed after hydration either by the cyclone or by careful sieving before bagging. The following molecular weights are useful in these calculations for limestones whose analyses are available:

Pure limestone (calcite)	$CaCO_3$	100.09
Pure dolomite	$CaCO_3.MgCO_3$	184.42
Quicklime	CaO	56.08
Magnesia	MgO	40.32
Hydrated lime	$Ca(OH)_2$	74.14
Magnesium hydrate	$Mg(OH)_2$	58.33

Quarrying and stone preparation
These have to produce raw material within certain size limits, depending on the size and type of kiln. As undersize material cannot be used in shaft kilns there will be losses in the quarry and at the crushing plant. This also applies when breaking rock with hand tools. Assuming it is not intended to grind the undersize material more finely so that it can be calcined in a separate kiln designed for the purpose, these losses will amount to around 50%, so the required reserves have to be increased by a factor of two to allow for this. This also assumes that the reserves identified only include the rock suitable for making the size of feed required and do not also comprise any shaley or soft marly layers that may be present in the formation.

The period of amortization
This is the number of years of economic life the kiln, hydrator and other plant are expected to have and over which their value can be depreciated for calculating return on investment, etc. The limestone must be in sufficient quantity to last out for that time otherwise part of the original capital value of the plant will have been wasted. The write-off period will vary greatly depending on the scale of the project and the standard of construction of the equipment. At one end of the scale, a small pit kiln which could be constructed in a few man-days would have very little investment cost and this could be written off after the kiln had been fired only a few times. Its depreciation period may be only six months. A traditional stone-built vertical shaft kiln, producing six to ten tones per day may be expected to last for ten years but will probably require major repairs after five years. Larger, modern kilns, electro-mechanically controlled and operated, usually imported and with outputs of 50 to say 200 tonnes per day would be depreciated over about twenty years. However, it would not be considered adequate to have raw material assured for only that time span. Big lime-works, like cement

factories, usually continue operating on a site for many more years. Once established, the infrastructure of access roads, offices, laboratory, railway siding and operating quarry as well as the nucleus of trained personnel with homes in the vicinity, all make it a more attractive proposition to continue longer if possible. Ideally then, reserves for 40 years or more are sought. Management of cement factories will apply to acquire raw materials to last for 100 years if they are known to exist close enough to the factory site.

As an example, suppose the project is for making 5000 tonnes per year (i.e. 15 tonnes per day for 330 days) of hydrated lime for a chemical use such as the sugar industry. This would mean finding a source of high calcium limestone, that is one at least 95% $CaCO_3$.

5,000 tonnes	hydrated lime a year
× 2	losses in calcination and cycloning partially offset by gain due to hydration
× 2	quarrying and stone preparation (crushing) losses
× 25	years of amortization.
= 500,000 tonnes	metric tonnes of economically exploitable resources in the deposit (classification R-1-E of the International Classification of Mineral Resources, see table on page 28, and broadly equivalent to 'proven reserves').

Such a quantity may be difficult to visualize until it has been converted into volume or, preferably, examples of thickness and area.

The *in situ* bulk density of limestone varies:

	kg/m³
pure dolomite	2840
pure calcite	2720
limestones	1900–2800
chalks	1500–2300
dolomite rocks	2600–2900

Taking 2350 kg/m³ for this example gives:

$$\frac{500,000 \text{ tonnes} \times 100 \text{ kg/tonne}}{2350 \text{ kg/m}^3}$$

= 213,000m³

In situ resources: quantities of economic interest for the next few decades.

R–1	R–2	R–3
Known deposits, reliable estimates.	Extensions of known deposits and newly discovered deposits, preliminary estimates.	Undiscovered deposits – tentative estimates.

R–1–E	R–1–S	R–2–E	R–2–S	R–3–G
Economically exploitable	Sub-economic	Economically exploitable	Sub-economic	Prospective geological potential

R–1–M
Marginally economic

Capital 'R' denotes resources in situ.
'r' would express recoverable resources in each category, e.g. r–1–E.

International classification of mineral resources (From *Economic Report No. 1, May 1979 of Energy and Mineral Development Branch*, Centre for Natural Resources, Energy and Transport, UN, NY 10017.)

One might suppose that quite a large area would have to be claimed to accommodate such a quantity of rock. In fact, though, even if there is just one metre layer of economically recoverable rock suitable for burning, the area of land required is only

$$\frac{213,000}{100 \times 100} = 21.3 \text{ hectares, or about 52 acres}$$

A 10m thickness of the deposit would mean a much smaller area: 2.13 hectares (5.25 acres).

As another example with a smaller output, there is a recently developed 3 tonne per day oil fired shaft kiln in Indonesia. Its amortization period might be, say, ten years and suppose the raw material is an argillaceous limestone for making a hydraulic lime:

3 t/d   output of quicklime
×330   operating days in a year
× 10   years depreciation
×100   factor for calcination loss of pure lime
  56   stone, i.e. 44%
× 75   factor for argillaceous limestone having
 100   say, 75% $CaCO_3$
× 2   quarrying and stone preparation losses
= 26,500 metric tonnes

Taking 2,500 kg/m³ in this case as the *in situ* bulk density of argillaceous limestone gives:

$$\frac{26,500 \times 100m^3}{25000} = 10,600m^3$$

In this case a one metre thick bed would need to extend only over an area of one hectare to be worth considering for the 3 tonnes per day kiln. Such small deposits are unlikely to be recorded on the geological map unless they form a chain of occurrences.

As transport costs rise so, increasingly, the smaller deposits of limestone will come to be utilized. They will be burnt in small kilns operating to supply the needs of the population in the nearby villages.

Scope of the project
Once the possible extent of the limestone resources has been found and claimed, the next stage is to discuss the project with geologists at the local office of the government's Geological Survey Department. Before going there, the lime project staff have to be clear what the objectives of the prospection are to be.

If the project is aimed at making lime for agricultural and building purposes, then the area to be prospected will be near those markets, probably somewhere within a radius of 20 to 100 km. A project to make higher quality lime, for a major industrial consumer such as a steel works or sugar refinery, will be likely to have a regional or national area of search. The economic area in which to find the raw material depends on the market location and the available freight transport systems and their cost.

As indicated previously, it is cheaper to transport the product, lime, rather than the raw material, limestone. Various local factors have to be considered. The existence of a navigable river, for example, between the raw material source and the market is a big advantage which many make it worth while to transport limestone. This is done through the Great Lakes to ports in Ohio and Minnesota, etc. In Paraguay, the country's principal industrial mineral resource is the high quality limestone at Vallemi. A cement factory is located by it and also some lime-kilns. Other kilns operators, though, send their limestone from Vallemi by barge down the Paraguay river to kilns located near to the capital, Asuncion.

The Geological Survey

This is usually a department of the Ministry of Natural Resources or Mines and Energy. Qualified geologists will be available to give general information about the rocks, such as limestones, available in a particular area. There may even be someone there who has done fieldwork in the area of interest and who can describe access routes, rock characteristics, thicknesses, extent and so on.

If the area has been mapped, copies of the 1:50,000 published geological map sheet should be available and often an accompanying memoir. Copies of each should be purchased, either at the Geological Survey or at the government bookshop or other sales outlet. The geologist may also be able to consult aerial photographs and satellite (Landsat) imagery of the area.

The memoir for the area, or separate reports, may include some chemical analyses of samples taken by the Geological Survey. Beyond this no further advice on the suitability of the rock for lime is likely to be available. This is because the behaviour of limestone in the kiln is so unpredictable that it is only determined by practical experience of lime-burning, either commercially or experimentally.

This general advice will usually be given free. The geologists will not normally be expected to assist with the prospecting in the field unless they are contracted to do so.

Topographic maps are needed for assisting in location of the area and marking the claim. Such maps will be purchased from the government's Land Survey or other cartographic department.

Prospecting and Mining Regulations

The Geological Survey will also be able to advise about any regulations existing for the control of mineral exploration and mining. There may be a separate Mines Department, with a Commissioner for Mines and Minerals, to whom a formal application for issue of, say, a non-exclusive prospecting licence for a particular district can be made and with whom a mining claim should later be registered. This would be the procedure to adopt for most of the smaller-scale lime-making projects, which might peg claims for up to about two hectares each. Larger projects, backed by more investment capital, would seek to obtain an exclusive prospecting licence and apply in due course for a mining licence covering the larger area such projects would expect to require. In 1981 the Malawi Government published its new regulations, under a Mines and Minerals Act, and these could well serve as a model for other countries at a similar stage of development.

Upon issue of a licence, the mining control body will notify both the applicant and the local district authorities. The prospector will report to them, as well as the land owner, as a matter of courtesy, before beginning any work in the area.

Field work
It will be invaluable to have the services of a professional geologist for at least part of the time in the field. Besides being able to recognize the rock type and its likely extent, the geologist should be able to advise on the effect of any overburden (the material covering the deposit) and other possible problems. The geologist can also advise on the method of working the deposit, including the best position to start the quarry operation. The initial reconnaissance will be followed by detailed investigation of selected areas. This includes sampling and field measurements, possibly trenching, followed later by a drilling programme for larger projects.

Before proceeding to the area, a route will be planned which takes advantage of higher ground, to provide an overall view of the country. This will not only be of the rocks but also the possible access routes, any railway or water transport systems, previous kiln sites, existing lime-kilns and any local laboratory facilities.

Among the items to be taken on the fieldwork are:

Geological hammer
Heavy hammer
Cold chisel
Hand lens ($\times$ 10)

Notebook and material data checklist cards
Sample bags and labels
Securely stoppered plastic bottle of dilute hydrochloric acid
Bottles of staining solutions
Magnetic compass, tape measure, etc.

The vehicle to be used will have four-wheel drive and preferably a long wheelbase. After reporting to the designated local authority office, the kilometer reading on the speedometer of the vehicle is noted, or returned to zero, and readings noted later at significant points such as possible quarry sites, the sample points, railway crossings, etc. The distances noted will help in the drawing of the sketch map of the area.

The purpose of the reconnaissance is to identify the limestone occurrences and choose the most promising places for detailed measurements of the exposed beds, possible sample points and likely siting of trenches to be dug. Some time should be taken in climbing to a vantage point, such as a hill, from where the maps can be oriented and the access, drainage pattern, changes in vegetation, etc. can be seen.

It has been said that time spent on reconnaissance is seldom wasted. This is particularly applicable when selecting a possible source of raw material for building material production, including lime, as well as choosing the site for the kiln or other plant. The ease, and thus the cost, with which the principal raw material, can be won will have a profound and lasting influence on the success, or failure, of the venture. The quality has to be the best available in that area. Too often, samples are taken at the first exposure encountered instead of waiting to see what lies over the hill.

Limestone is usually a fairly hard and durable rock so that it forms features on the landscape such as a line of hills, or a scarp or cliffs. As it is slightly soluble in carbonated water, there may be caves, or swallow holes, with stalactites and stalagmites. The soil cover is usually rather thin. The soil may be somewhat acid where there is leaching over a porous limestone formation. The colour of the rock varies from medium to light grey, buff, cream and nearly white.

Calcite has a density of 2.71 and its hardness is 3, on Moh's scale. So it is harder than gypsum (2) and a fingernail (2.5) will not scratch it. It will be scratched by a copper coin (5) and window glass (6).

There is no problem in distinguishing it from quartz as the latter is much harder (7).

A chip of calcitic limestone or marble will effervesce vigorously in dilute hydrochloric acid and leave little or no residue. Dolomitic limestone will fizz slowly while dolomite will react only if the acid is heated. Calcite and dolomite can also be distinguished by staining tests. A 30% solution of aluminium nitrate ($Al(NO_3)_3$) will stain calcite blue but not affect dolomite. If Alizarin Red is sprayed onto a surface that has been washed with dilute hydrochloric acid the dolomite is unaffected whilst the calcite is stained pink or red. This affords a simple method in the field of comparing the dolomitic proportion of limestones.

The following work is carried out in the detailed field investigation:

Exposures of limestone in cuttings, stream sections and old quarries are measured to determine the proportion of rock that appears suitable, in relation to the whole formation. Its thickness is noted. The dip angle and direction are noted and also the thickness of overburden. About 100kg of this rock is collected, in fresh lumps both for firing trials, as pieces around 6–9cm size, and for chemical analysis.

The visible line of continuation of the limestone between outcrops is noted on the map, or on a sketch map. the soil may be thin and support a sparser vegetation than adjacent lithologies. This and the topography may indicate possible extensions of the limestone.

A sketch map of the deposit is drawn and will include old quarry sites, access roads, distances and directions to features such as hilltops, villages, etc, the angle of dip and its direction, the slope of the ground, surface drainage routes, etc.

Note is made of any suitable location for the kiln, especially where the slope of the ground will make loading a much simpler operation.

Shallow trenches, up to about 0.5m deep, are dug across the strike of the beds using local labour. These trenches will confirm the presence of the rock and provide further sample points for taking fresh, unweathered material.

A reliable estimate of the 'visible reserves' (the economically exploitable resources) is made by taking conservative figures for the average thickness and the visible extent of the deposit.

Any other points of possible future significance, such as

inadequate bridges over access roads, are noted.

An example of a two-sided card bearing a check list of data about the deposit is shown below.

DEPOSIT DATA		
Region/District	Refs:	Rock/Mineral

Location
  Map Sheet
  Grid Ref.
  Elevation
	Place
	
	Route
		Owners/Tenants/
		Licensee

Size
  Extent
  Thickness
	Resources (UN)	

Character
  Form
  Bedding/Joints
  Dip
	Overburden
	Drainage/Water Table	

Other Information
  Truck Access
  Electricity
  Water
	Present Utilization	Date

MATERIAL DATA		
Region/District	Samples No. Refs:	Position
Analyses/Testing Refs:		
Industrial Potential		Decision/Further Work

The reverse side of the card is for subsequent recording of the laboratory analyses and mineralogical examination.

The report of the field work should include:

1 The location of the likely suitable deposit(s), including the distance from main the market.
2 The visible reserves and possible additional reserves (economically exploitable resources) (R-1-E) and economically exploitable extensions of the deposit, preliminary estimate (R-2-E).
3 The likely quarry and kiln positions.
4 The ease of access and quarrying and any special problems.
5 The samples taken and any conclusions reached from lithology and field tests, including any small burning trials that were possible, about the likely performance of the material for lime making.
6 Any further work required, such as drilling.

Drilling, which here means taking rock core samples with a diamond tipped core bit, is expensive. The cost varies widely from place to place and there may be no drilling contractors in a region. There is usually inadequate information about the raw material supply, such as the reserves and their variability. Proving the raw material, by drilling, surveying and analysis, can be expected to consume 2–3% of the projected budget, often 5% and sometimes as much as 10%. Even drilling one hole down to 50m and moving the rig to and from the location could cost US$5000, say, or more in areas of difficult access. So it is obvious that drilling is not going to be feasible for any small-scale lime projects of up to about 10 tonnes per day and costing US$100,000 to 200,000. When the total investment is say US$500,000 or more, there is a greater need to take out an 'insurance' by checking the reserves more thoroughly. At 5% this means the possibility of spending perhaps US$25,000 on drilling holes to back up the estimates made from the examination of exposure, trenches, etc., on the surface.

The number, spacing and depth of the holes can be kept to a minimum by knowing in advance the planned production capacity of the kiln, its economic life and, hence the volume of rock to be proved. Proving more than that may be an unnecessary expenditure of investment capital.

3.4 Laboratory work

On large projects costing US$1 million or more the consultant or kiln manufacturer will sometimes arrange for the raw materials to be analysed and tested, often in an overseas laboratory. This need not be the case for the small projects of up to 20 tonnes per day and it may be inappropriate to include extensive sample analysis if the market is to be for simple building uses. It is more practical and as quick to test fire the limestone and to slake the product. The degree of whiteness for limewash and the plasticity and hardening properties for mortar and rendering are then readily compared between samples. Appropriate tests for the particular needs of each small scale project will often be as simple as this. In many cases it is wrong to look for the more detailed and precise data which our scientific training misleads us to expect.

Chemical analyses are useful mainly for comparing deposits and for indicating the possible degree of variation within a deposit. These analyses can be done by classical wet methods in the laboratory of a cement works or of a geolocial survey. For instance:

Minimum Analysis
    CaO
    MgO
    Insoluble residue
    Loss-on-ignition
Additions for 'Full' Analysis
    SiO_2
    Al_2O_3
    Fe_2O_3
    P_2O_5
    $S--$
    SO_4--
    Na_2O
    K_2O
    TiO_2

Loss-on-ignition (LOI) is determined by drying a part of the ground up sample at 110°C and then noting the loss in weight after firing at 1000°C until there is no further weight loss. Most of the weight loss will be due to carbon dioxide driven off from the carbonate. The total carbonate content can then be deduced by comparison with the figures for CaO and MgO. This and the separate values for $CaCO_3$ and $MgCO_3$ immediately give a rank to the limestone as a raw material for industrial purposes, as shown earlier. The LOI though, will be increased by any carbonaceous material present unless the ignition is carried out in steps to distinguish it.

The following data will assist in interpreting the results presented by the analyst:

Analysis	Carbonate
56% CaO and 44% LOI	100% pure $CaCO_3$ = pure limestone
CaO > 53.2%	$CaCO_3$ > 95% = high calcium limestone
30.4% Cao + 21.9% MgO + LOI	100% $CaCO_3$ $MgCo_3$ = Pure dolomite
2.4% < MgO < 19% + CaO + LOI	Magnesian or dolomitic limestone
10% < SiO_2 and Al_2O_3 < 25% and remainder is CaO and LOI	Limestone which might yield a hydraulic or semihydraulic lime.

Even with these results, there is no certainty about how the stone will perform in, say, a shaft kiln. It is still essential to carry out trial firings to see if the stone maintains itself as lumps during calcination to quicklime. It is very instructive if small-scale firing trials can be arranged, perhaps in the area of the deposits. This will become more necessary as more small kiln projects are planned to serve local populations where supply from the main production centres is uneconomical. The entrepreneur or project officer with limited financial resources will find that it is surprisingly simple to arrange such firing and slaking trials.

3.5 Small-scale limestone firing and lime slaking trials
The purpose of the small-scale firing trial is to see if the stone is likely to burn well in a continuously operated shaft kiln, or whether it will have to be burned some other way. The lime slaking trial will

A kiln for small-scale firing trials.

indicate the reactivity and apparent quality of the lime that the stone can make.

The drawing shows an example of a small batch kiln, built in Malawi. There it was used to assess whether a coarsely crystalline marble would decrepitate or maintain a lump form and so be likely to fire more efficiently in shaft kilns. The capacity of this kiln was about 250kg of raw material and it was fired with wood. Some coal could also have been burned.

An even smaller kiln, of about $0.04m^3$ capacity (just over $1ft^3$) to fire 50 to 75kg of rock, is useful for trials during field work. It can be built in about three hours using up to 300 fired clay bricks laid with mud mortar. Stone or sun dried bricks could also be used. One built in northern Malawi made use of scrap drill rods and old steel plate and had a disused water pump casing for a chimney — it was used to investigate sedimentary argillaceous limestones. From the start of construction to preparing the stone into equidimensional (cubical) lumps around 6 to 8cm size, cutting up the firewood and loading the kiln ready for firing took 5.5 hours. After 10 hours firing and 24 hours to cool down the kiln was opened to remove and inspect the quicklime.

In both of these trials the behaviour of the limestone was quite the opposite of what is normally expected and the test results were very helpful. The sedimentary limestone that was fine grained, hard and of uniform porosity, tended to explode with considerable force so that many smaller fragments were produced. The coarsely crystalline metamorphic limestone, or marble, showed no tendency to decrepitate in the kiln.

A shallow brick-lined slaking pit prepared near the kiln is used for observing the slaking behaviour, i.e. the reactivity of the quicklime, and to make lime for mortar and whitewash assessment. Reactivity of quicklime also depends on firing conditions but using wood there is little risk of rendering the lime unreactive by overburning.

Clayey soil is ideal for making the mortar to build the kiln. It is also used to 'scove' or render the exterior to reduce draughts as well as to cover the loose bricks used to seal up the front of the kiln. Cement or lime mortars are unnecessary and will crack.

If a limestone retains its form when fired in such a kiln and has adequate physical strength to resist abrasion, the next step will probably be to construct a small stone or brick built mixed feed vertical shaft kiln at the limestone deposit. Using that it will be possible to prove whether the raw material is suitable for continuous and thus more fuel-efficient lime production in conventional shaft kilns.

3.6 The raw material report

The raw material report must be presented to the project manager to show whether the proposed lime-making project is technically viable. It must include the following findings:

The grade of lime that can be made.

A reliable estimate for the economically exploitable resources, in metric tonnes. Also any preliminary estimates for the economically exploitable resources from extensions of the deposit or newly found deposits. Hence, knowing the planned production capacity of the project, the number of years of raw material resources available.

The cost of getting each metric tonne of prepared limestone into the process.

An indication of the suitability of the raw material for feeding into the more fuel efficient types of kiln, i.e. resistance to decrepitation and abrasion.

The location of the deposit, its relation to the intended market

and alternative positions for siting the kiln. Any special problems, such as one access, quarrying, etc., that may not be
reflected in the raw material cost.

The types, availability and cost of fuel in the area.

There will be other information that will need to be known about
the proposed area for the lime plant, and much of it can be learned
during the field work if it is not already known. Examples are:

the selling price of lime in the area

communications such as transport routes for the lime to the
principal consumers

available manpower

water, electricity and other services

workshops in the area

the industrial development planning for that area (will have been
checked at the beginning)

the ownership of the land

any local pozzolanic materials (volcanic ash and pumice, rice
husk ash, diatomaceous materials, etc.).

The availability of suitable pozzolanic material could make it
possible to promote the production of lime/pozzolana cement,
block making and other building products. Pozzolans are materials
which can be mixed with lime to form special cements.

The type of lime, or the quality that can be made, will depend
primarily on the chemical composition of the limestone. However,
a proportion of the impurities can be removed during the sieving or
cycloning of the hydrate.

The traditional way of listing the *in situ* resources has been as
proven reserves, probable reserves and possible reserves. The
number of years that the project can operate is based on the 'proven
reserves'.

The calculation of the cost of raw material takes into account
factors such as:

ground rent, per annual tonne

royalty per tonne of limestone

capital equipment charges, e.g. rental or interest payment on drill, shovel, trucks, crushing and grading plant, pumps, if any, etc.

fuel and maintenance

labour and supervision

quarry and crushing plant

insurance

other consumable items such as explosives and protective clothing.

Before deciding whether to use mechanized plant or labour-intensive methods for extracting and breaking the rock, the two systems should be compared. It may not be practical to operate and maintain modern equipment if there is no skilled labour or specialized workshops or if only small tonnages are to be produced. Spare parts may be very expensive and take months to be delivered. During this time the operation stands idle. In areas of low wage rates it may work out to be entirely feasible and appropriate, economically, socially and environmentally to employ many people, perhaps several family groups, in the quarrying, hand selection, breaking of the rock and delivery to the kiln.

The suitability of the raw material for shaft kiln operation will be based partly on the physical characteristics of the rock, but primarily on the evidence of firing trials.

To show the quarry and kiln locations, a sketch map should be prepared, to scale, and this will include:

roads and tracks

distances and directions to nearest towns

north point

outline of the deposit

possible initial cut to open the quarry

an indication of other benches to cut to keep the bench height low

possible kiln locations

hydration plant

streams

distance to railway or navigable river

power line etc.

On large projects it will be necessary to make provision for disposal of the undersize rock. This may be sold as:

graded aggregate for concrete

bitumen coated road aggregate

chippings

finely ground powder limestone for 'agricultural lime'

finely ground powder as a filler.

Availability or otherwise of fuel, whether it be wood, coal, oil, gas, agricultural waste, etc., could be the deciding factor in choosing between two locations for a lime-making project. Increasingly, the provision of nearby land for plantations of fast growing trees to ensure adequate fuel supply has to be considered in the development of lime as well as brick and tile making industries.

CHAPTER 4
Fuels for Lime-kilns

4.1 Introduction
The choice of an appropriate fuel is of very great importance to the success of a lime-works. Although most fuels can be used in some way or another it is always necessary to strike the right balance between simplicity, fuel cost and efficiency.

The traditional fuels are wood, coal and sometimes peat. In many ways wood is one of the best fuels and when other fuels are used attempts are made to imitate the best flame characteristics of wood.

The most expensive coals burn best in lime-kilns and some lime-burners keep down costs by burning high quality coals in the least popular size gradings.

Oils and gases can be efficient and convenient but the balance is offset by the need for the more sophisticated types of kiln with higher capital and maintenance costs.

There are special techniques for burning the more difficult (and cheaper) fuels. Some of these will be helpful for small scale lime-burning although in all cases a little further development seems to be needed.

If renewable fuels such as wood and biomass are to be used, then replanting on an appropriate scale is essential and the new crops must be properly maintained.

4.2 Wood
Wood was the original fuel used for lime-burning and wood firing still produces some of the best quality limes. It can be used in batch fired kilns or in shaft kilns. Its great advantage is that it burns with long, even flames of mild, 'moist' heat, and requires only natural draught. These long flames enable the heat to penetrate into the mass of limestone and create a broad burning zone. This makes good use of the kiln capacity and gives a uniform calcination.

Steam is generated from the wood as it burns and this helps to lower the temperature needed for dissociation. The low flame temperature make it almost impossible to overburn lime using wood.

Unfortunately, however, the great popularity of wood as a fuel for domestic purposes as well as for lime- and brick-burning has led to the large-scale destruction of natural forests. In England, for example, this problem had become serious by the thirteenth century and coals had to be mined and transported over long distances for lime-burning. Lime manufacturers had to make a number of modifications to the method of burning and to their kilns to achieve lime as good as that burned with wood. Examples of this were the control of the draught and the humidity in the burning zone. The history of lime-kiln design is, to a certain extent, the history of the development of alternative fuels.

Properties of wood
The following properties of wood are relevant:
The *chemical composition* of air-dried hardwood is approximately:

Carbon	(C)	40%
Oxygen	(O)	34%
Water	(H_2O)	20%
Hydrogen	(H)	5%
	Other	1%
		100%

The *density* of fuelwood can vary considerably according to the species, but the solid density of dry wood is about 0.6 – 0.9 tonnes/m^3. The bulk, or stacked, density is about two thirds of this.

The *calorific value* is roughly proportional to the density, and is very approximately 12900 – 17000 kJ/kg for air-dried woods and 17000 – 21000 kJ/kg for oven-dried woods. The value is enhanced by about 20% if the wood contains resins, oils or gums. Green (freshly cut) wood has only about half the burning efficiency of seasoned (dry) wood, because much of the energy content is wasted in evaporating excess moisture.

The *fuel consumption* (firewood consumed per unit weight of lime) varies considerably according to the kind of wood and type of kiln used. Experience in different areas has given limestone/

fuelwood ratios ranging from 3:1 to 6:1 for batch kilns. Semi-continuous kilns need about 20% less fuel. Recent experience in East Africa with a 1 tonne per day semi-continuous vertical shaft kiln using pine fuelwood showed a ratio of limestone/fuelwood/hydrated lime of (2.5–3.0):(0.7–0.8):1, which is a limestone/fuel ratio of about 3.5–4.0, that is about 250–280kg of wood per tonne of limestone.

Wood stacked carefully for measurement. The picture represents 250 kg wood and one tonne of limestone.

The *rate of burning* is an important property of fuelwood. Where fast burning woods like coconut have been used they have been mixed with a slower burning wood. The former maintains the combustion of the latter.

Harvesting, preparation and use of wood
The harvesting method will have to be selected according to local conditions, such as soil, terrain, plant patterns, tree species, size and the availability of capital.

Since the wood required for lime-burning will be used in relatively small pieces, it will normally be possible to harvest economically with lightweight equipment. Felling is best done with a bow saw but a chain saw could be used, particularly if trees are larger than 200mm in diameter. General advice on harvesting is given in the book *Appropriate Technology in Philippine Forestry* (see Chapter 12).

The wood may be cut into suitable sizes in the forest or at the lime-works. Transport can be by pack animal, animal or tractor-drawn trailer or lorry. The wood supply should be as close to the kiln as possible to avoid heavy transport costs. As green firewood gives out only half the heat available from seasoned wood, firewood must always be dried before use. Logs should be split if their diameter is greater than 150mm and they should be stored for about six months. This means that enough working capital must be available to accumulate six months' supply of wood and appropriate credit facilities must be available.

The size of wood pieces required will depend on the species and kiln type. Pieces 500mm long and split from 150mm diameter logs were used successfully in a 1.2m diameter mixed feed shaft kiln in Tanzania. In other types of kilns lengths of 1–2m have been used. These figures are given for guidance only and experiments should be carried out locally to determine the best size.

If the pieces have become very dry, they should be immersed in water for a few seconds before they are loaded into the kiln.

When a semi-continuous, mixed feed kiln is loaded, the lowest layers should contain more than the normal proportion of wood. In the example in Tanzania the kiln charge was built up as shown in the following diagram:

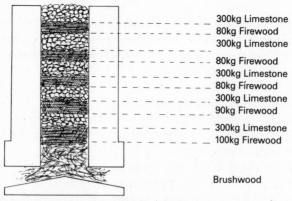

300kg Limestone
80kg Firewood
300kg Limestone
80kg Firewood
300kg Limestone
80kg Firewood
300kg Limestone
90kg Firewood
300kg Limestone
100kg Firewood

Brushwood

The initial charge for mixed feed continuous firing.

The limestone is broken into fist-size lumps. During the operation of the kiln alternate additional layers of firewood and limestone are fed into the top of the kiln, and cooled calcined lumps of quicklime are being withdrawn through the discharging tunnels at the base. This charging and drawing is repeated throughout the period of operation of the kiln. For maximum efficiency the cycle should be repeated as often as possible, but it is difficult to draw lime more often than three or possibly four times each day without working at night.

Fuelwood plantations

Even small lime-kilns consume very substantial amounts of fuel. For each tonne of hydrated lime produced about $2m^3$ bulk of wood might be needed. It is therefore essential to consider how the long term supply of the wood can be ensured if wood is to be used for lime-burning. There are very few areas where the regrowth rate of natural vegetation would replace that cut down and so a fuelwood plantation is likely to be needed.

Purpose grown plantations have the advantage that the sites can be selected to make eventual harvesting simple and the trees can be grown to the most convenient diameter to avoid the need for splitting.

The planning of such plantations is complex and requires specialized knowledge. However, as a very rough guide to the size of plantation required, two recent studies on fuelwood for lime-burning in Ol Donyo Sambu and Vanuatu have both shown that about twelve to sixteen hectares of plantation would be needed for each tonne of hydrated lime to be produced each day. The size of plantations which would be needed depends on such factors as the species type, altitude, rainfall, kiln output and the efficiency of management of the plantation.

The table overleaf shows some representative data for common species types: further information is included in the book *Firewood Crops* by the National Academy of Sciences (see Chapter 12).

Local species should always be tried before new ones to avoid possible disturbance of ecology.

For further information the district or regional forestry departments or the national forestry institutions should be approached. If they have only limited experience of energy plantations specialist advice may be obtained from international forestry organizations such as The Food and Agriculture Organization of the United Nations, (FAO), Rome; The International Council for Research in Agroforestry (ICRAF) Nairobi; Appropriate Technology for Forestry (ATF), London, which is a partnership of the Intermediate

Firewoods

Species	Required Rainfall (mm/year)	Altitude (m)	Specific gravity	Calorific value (oven dry basis) (kcal/kg)†	Yield (m²/ha/year)	Coppice rotation (years)
Acacia Auriculiformis*	1500–1800	0–600	0.60–0.75	4800–4900	10–20	10–15
Leucena Leuconephala*	600–1700	0–500	0.70–0.80	4200–4600	15–30	8–15
Azadirachta Indica	450–1500	50–1500	0.56–0.85	4900	5–20	8–15
Cassia Siamea*	700–1100	0–500	0.60–0.80	(4600–4800)	10–15	8–10
Prosopsis Ceneraria*	75–1850	0–600	0.60–0.80	4900	5–30	8–15
Eucalyptus Camaldulensis	500–1100	30–600	0.70–0.85	4700–4800	5–30	7–15
Eucalyptus Tereticornis	500–1500	0–1000	0.75–0.85	4700–4800	5–25	8–12
Alibizzia Lebbek*	500–1000	0–4000	0.55–0.65	(4400–4700)	15–25	15–20
Eucalyptus Grandis	500–2000	1000–4000	0.50–0.65	(4400–4700)	20–30	15–20
Gmelina Arborea	1000–2300	0–800	0.40–0.50	(4200–4500)	15–30	12–20

* These species are from the family Leguminosae, normally forming root nodules with nitrogen fixing bacteria and therefore have useful soil improved capacity.

† Figures in brackets are estimates as no reliable measurements are available.

Technology Development Group, The Commonwealth Forestry Institute and The International Forest Science Consultancy.

Charcoal

Charcoal is formed by burning wood under conditions where it has only a restricted supply of oxygen. The final product contains more than 80% carbon and is a very efficient fuel, which burns with very little smoke. Its calorific value is about double that of dry firewood. It has a bulk density 40% lower than that of dry firewood and this gives charcoal great advantages in long-distance (greater than 100km) transport, and also in storage and trade. A study called *Charcoal in the Energy Crisis of the Developing World* investigates these advantages (see Chapter 12, ref. 33).

Sometimes, as in Ethiopia, the charcoal is soaked in water before it is fed into the kiln.

Charcoal has been used in the manufacture of Portland cement, for example in Uganda as described by Earl (28). Information is lacking on its use in lime manufacture. However, although using charcoal is certainly more efficient than traditional wood-burning methods, it is not as efficient as an *efficient* wood-burning process such as a well-designed lime kiln. Also, the production of lime using charcoal would mean an awkward two-fold process, first preparing the charcoal and then the lime. For reasons described above, lime burned with charcoal would not be as well burned as lime burned with wood. It is therefore unlikely that the use of charcoal would improve the efficiency of wood-burning lime-kilns.

4.3 Coals and coke

Coal is found in every continent and is the most abundant of the fossil fuels. It was formed in early geological times from decaying plants and this would suggest that there may still be large resources undiscovered. It has been used in lime-burning for over 700 years and will continue to be an important fuel.

Coke can be made from coal by removing the volatile content for other uses to leave a fuel with a very high carbon content. Coke has been one of the most successful fuels for lime-burning as the way it burns is very suitable for use in kilns where the fuel and limestones are mixed together.

Dryden's classification of coals

	Carbon	Volatiles	Hydrogen	Calorific value kcal per kg	Rank
1. Anthracites	92 – 95%	10 – 3.5%	4.0 – 2.9%	8870 – 8353	Highest
2. Bituminous coals	75 – 92%	50 – 11%	5.6 – 4.0%	6985 – 8870	↓
3. Brown coals and lignites	60 – 75%	60 – 45%	5.5 – 4.5%	6653 – 7207	
4. Peats	45 – 60%	75 – 45%	6.8 – 3.5%	4157 – 5322	Lowest

Properties of coal

There is a wide range of coals which can be classified in various ways. Dryden's classification has four main divisions:

The divisions in this table show the successive stages in the metamorphosis from plant materials towards almost pure carbon. The changes were brought about by the action of high temperature and pressure over long periods of time. The word 'rank' is used to show how far this change has advanced in a particular sample of coal. Thus peat, which has been subjected to only normal temperatures and very light pressure is the lowest rank coal whilst anthracite, which occurs only in deep mines, is the highest rank coal.

A further important characteristic of coals is their volatile content. This is the measure of the hydrocarbons which can be driven off as vapour by heating the coal. The volatile content is greatest in the lower ranking coals and decreases as the rank increases. Coals known widely as 'long flame coals' and 'cooking coals' are found in the middle ranking bituminous class. It is their volatile content which burns to give the long flames. The ability to form 'coherent coke' is dependent on both the volatile content and on moisture.

The related properties of carbon content, hydrogen content, calorific value, volatile content and rank are shown on the Sayler chart which, in a simplified form, is shown as Appendix 9. It gives a useful way of predicting the value of any available coal for a particular purpose. The shaded area between the two curves includes most of the commercially important hard coals. If any two of the properties are known the others can be read off the chart. The chart shows the four categories of Dryden's classification and some of the other commonly used descriptions.

There are several difficulties in using coal as a fuel for lime-burning but with care it can be used to make good lime. The best coals for a lime-kiln would have a high carbon content but the higher rank coals are generally too expensive to be economic. Coals for economic lime-burning might have a carbon content of between say 60% and 76%.

The calorific value, that is the amount of energy available in a unit weight of the fuel, is highest for the highest rank coals and lower for the lower rank coals. The 'proximate analyses' of two such coals are shown in the following table:

If this information is transferred to the Sayler Chart it can be seen that both samples fall outside the shaded area and so would not be

	Example A (Lower rank)	Example B (Higher rank)
Carbon	60.6%	76.6%
Total Volatiles	21.8%	14.2%
Ash	8.3%	6.3%
Moisture	9.3%	2.8%
Gross Calorific Value	6110 kcal/kg	7545 kcal/kg
Nett Calorific Value	5842 kcal/kg	7141 kcal/kg

Proximate analysis of two coals showing the proportion by weight of important constituents.

commercially important varieties. For this reason the prices should be relatively low.

Coal has often been used for lime-burning in kilns with equipment designed to make producer gas from the coal on the site. This method of use is described below in Section 4.4 along with other gaseous fuels. The use of the Sayler Chart will be considered again in that section.

Properties of coke
Coke is made by heating a suitable grade of coal strongly in a closed vessel, a retort or a coke oven. The closed vessel prevents oxygen from reaching the coal to burn it but allows the volatile content of the coal to be driven off. The hot coke is immediately quenched in water to prevent ignition when it is removed from the vessel.

Different grades of coal will form different cokes. The expensive 'metallurgical' grades have a silvery appearance and a hard sponge-like texture. The cheaper grades, which are more easily obtainable, are softer with a less obviously open texture and are dull grey in colour.

All cokes are suitable fuels for mixed feed shaft kilns and the lower grade cokes are usually better value than the purer cokes. A proximate analysis of good coke is shown below. A metallurgical grade would have a higher proportion of carbon and less ash.

Coke is a manufactured product which will not be available in all places. Its porous structure allows it to soak up a lot of water and if stored for long periods in wet conditions it will deteriorate.

	Composition by weight
Carbon	85%
Total Volatiles	4.3%
Ash	10.7%
Moisture	0.8%
Calorific Value	6883 kcal/kg

Proximate analysis of a good grade of ordinary coke.

How coal and coke are used for lime-burning

To understand how coal burns in a lime-kiln it is helpful to consider first how it burns on a hearth or open grate.

Some kindling, such as dry wood, must be used to set a piece of coal alight. The kindling must burn for some time before the first small flame appears from the coal. The coal starts to burn slowly and smokily, but if it is of reasonable quality a good bright flame will develop even if it remains rather smoky. This is the first of the two stages of the burning.

For some time this bright flame will burn above a glowing bed of coal but the fire will then begin to die down. It must be stirred up from time to time to maintain a clear flame and new coal must be added. If no further coal is added the fire will slow down until it just glows even though it still gives off a good heat. This is the second stage of the burning. When this burns out only ash is left.

During the first, flaming, stage the volatiles were being driven off and when the temperature was high enough they burned with a good long flame. The main component of the volatiles was hydrogen. When all the volatiles had been driven off only carbon and impurities remained. The carbon burned with just a rare flicker of flame. Coal burning on an open hearth will give off some smoke until almost the end of the flaming stage. This is because air to provide oxygen for combustion cannot reach the volatiles before the temperature drops below their ignition point. The smoke is the unburnt fuel escaping.

Coal can be burned completely in a furnace such as those used in old steam locomotives or small boilers. In these most of the air needed for combustion is drawn up through the bed of burning coal. But in addition to this there is an opening above the fire level

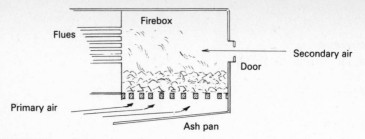

Locomotive firebox.

through which more air can enter. This secondary air is drawn across the top of the fire and helps to achieve complete combustion of all the volatiles with less smoke than is usual over a simple hearth.

It is difficult to see the inside of a lime-kiln even when inspection holes are provided. Our knowledge of how coal and coke burn in the kiln is developed from what we know of the combustion already described in an open grate and in a firebox. The important point is that secondary air is needed above the coal bed if the volatiles are to be fully burned.

This diagram shows a section through a mixed feed shaft kiln. The general features of the kiln are described in Chapter 5 and only the interior is shown here. The kiln is designed to work continuously for many months. Fuel and limestone are both introduced at the top and quicklime is drawn from the base.

The diagram shows four zones and whilst they are divided here by firm lines, in practice the functions merge into one another. The top zone is for storage to ensure that there is always sufficient material in the preheating and drying zone below it. Half way up the kiln is the burning zone where the temperatures are high enough for the coal to burn. Below this level is the cooling zone where the lime is cooled by the incoming air. All the coal should have burned by the time the material has reached this level in the kiln.

When the kiln is first started kindling is placed beneath the fire bars. Alternate layers of fuel and limestone are carefully placed for the full height of the kiln. Sometimes additional easily burnt materials are used at the bottom to help start the fire. The kindling is lit and the first layer of coal slowly catches alight. All the air for combustion passes up through the fire bars and at first the fire is very smoky. As the heat builds up combustion improves. The hot air rises up through the kiln and pulls in new air in the same way that

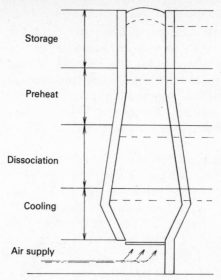

Zones before drawing

Storage

Preheat

Dissociation

Cooling

Air supply

Mixed feed shaft kiln.

a chimney pulls air through a furnace. The heat sets the next layer
alight, partly because the draught has made more air available and
partly because the air reaching the fuel here is already hot and so the
volatiles are not so severely cooled by their combustion air.

The fire spreads gradually up through the kiln and if no further
action is taken it burns right through to the top and dies out. This is
the technique used in intermittently fired mixed feed shaft kilns
described in Section 5.2 but to operate the kiln for continuous
production the lime-burner has to judge when the fire has worked
its way up to the intended burning zone. He then draws out some of
the quicklime from the base of the kiln and the whole charge falls
downwards. As it falls the coals and limestone become intermixed
and the charge loses its strictly layered form. New materials are
added at the top.

The lime is then drawn from the base of the kiln at regular
intervals and the lime-burner judges how much lime must be taken
out to keep the fire in the intended burning zone.

When the charge is first built the layers of limestone weigh about
four times the weight of the layers of good coal but when the kiln is

working in steady production the proportion of stone can be increased to five times or even more for a well-designed kiln.

If this operation is compared with the way in which coal can be burnt on an open hearth or in an improved furnace it is clear that the shaft kiln does not provide any secondary air above the burning coals to burn the volatiles fully. For this reason fuels with low volatile content are technically more efficient in shaft kilns. In the previous section the composition of coke was described and it was seen that the volatile content was driven off the coking coals for other profitable uses to leave a product — coke —with a very high carbon content.

When coke is used in mixed feed shaft kilns there is a further advantage. Cokes will not burn until the temperature reaches 800°C which is only a little below the dissociation temperature of the carbonates. This allows a more even distribution of temperature in the dissociation zone in the kiln with less chance of spoiling the lime through 'hot spots' which cause overburning. There should also be less unburnt fuel than is usual with coal and these factors combine to give greater fuel efficiency and better quality lime.

A charge of coke is harder to ignite than a charge of coal and some lime-burners will mix coal and coke in equal portions when the charge is first built and reduce the proportion of coal over the first five days of burning. A few English lime-burners used to add a little coal even during normal production as the flames were thought to keep the coke well alight.

4.4 Liquid and gaseous fuels

In many ways liquid and gaseous fuels are easier to handle than solid fuels. The movements can be achieved with simple pumps and pipes instead of the more troublesome conveyors or hand labour. They also burn with no ash and this can be seen as an advantage either for the preparation of an uncontaminated lime or to avoid any need to handle and dispose of unwanted ash.

Though coal is often cheaper in some countries it may be unobtainable in others, and there liquid and gaseous fuels must be considered even for the smallest scale of lime-burning. The precautions which are needed when burning with oils and gases seem to be more worth while on a large scale than on a small scale and the construction of the plant would nearly always be handled by a specialist supplier. For this reason no attempt is made to give any detailed specification of equipment.

Oils are found underground and are transported by pipeline or tanker. Natural gas is delivered by pipe whilst certain petroleum gases produced during the stages of oil refinery are compressed into liquids (LPGs) for convenient transport. They are then re-vaporized for use.

Oils

The mineral oils used for fuels are usually found in sedimentary rock formations. They were made by the metamorphosis of organic materials and the character varies from one deposit to another. The oil is extracted through wells and transported by pipeline or tanker to refineries.

The oil taken from the wells is called crude oil and is a mixture of many hydrocarbon chemicals. It is processed in the refineries to produce a range of products chosen to satisfy the market needs. These refinery products will include petrol and fuel oils and each grade will vary slightly from place to place. The data in the appendix are for typical examples of five widely available grades.

The two main divisions of these refined oils are light distillates and heavy industrial fuel oils. Within each division there are several grades:

LIGHT DISTILLATES

Kerosene: The lightest grade usually available. It is widely known by the incorrect name 'paraffin'. It is sometimes used for lighting up lime-kilns but would be too expensive to use as the main fuel for a kiln.

Gas Oil: Known, when used for heating, as '35 secs oil' and, when used for diesel engines, as Derv. An easily handled fuel but rather expensive for general use.

HEAVY INDUSTRIAL FUEL OILS

Light Fuel Oil: A general purpose fuel oil which flows well and is only likely to need preheating in cold climates.

Medium Fuel Oil: Another general purpose fuel oil but less easy to handle than the light fuel oil. It will need to be preheated for combustion and in cold climates the fuel stores will also need to be heated.

Heavy Fuel Oil: Also known as residual fuel oil. Heating is

essential for storage, outflow and combustion. It can thus only be used for large scale operations.

The lighter grades of fuel oil have rather higher calorific values, are less viscous and are cleaner. The extra difficulty in handling heavier oils should be reflected in their price. A very high sulphur content can be a disadvantage for lime-burning as the sulphur dioxide gas which is produced will contaminate the lime.

Some physical and chemical properties of typical samples are set out in Appendix 11.

Liquified petroleum gases (LPGs)

Liquified petroleum gases, which are nearly always known as LPGs, are produced from crude oil at oil refineries. The two important LPG fuels are propane (C_3H_8) and butane (C_4H_{10}). They are members of the same alkane (or paraffin) family as methane but unlike methane both propane and butane can easily be liquified under moderate pressure at normal temperatures.

Both propane and butane are supplied for small scale use in small portable cylinders, but butane is usually favoured for large-scale use. It can be delivered in road tankers for bulk storage at the users premises. In its liquid phase butane is denser than propane and it also has a higher calorific value. Both fuels are transported and stored as liquid but re-vaporized before burning.

LPGs are clean fuels — that is they contain no more than a trace of sulphur and so they cause very little sulphurous pollution when they are burnt. When they are used for lime-burning the lime is not contaminated.

Although LPGs are not poisonous, care is still essential. The gases are heavier than air and can collect in confined spaces. Both asphyxiation and explosion are real dangers and expert advice should be sought from the supplier.

Specialized storage and handling equipment is readily available, although rather expensive. Butane is only a commercially economic fuel for lime-burning where supplies are available near the works, where continuity of supply is expected, and where the high quality of the fuel can be justified.

Properties of typical examples of propane and butane are tabulated in Appendix 12.

Natural gas

Natural gas occurs in similar locations to oil deposits. It is a mixture of methane (CH_4) and small amounts of other alkane (paraffin) gases with a trace of nitrogen.

Although all of the present sources are deep wells it is thought that there may be other types of deposit with very large hidden reserves.

The gas is supplied to users through pipelines and measured by meters at the point of delivery. A typical delivery pressure which would be high enough to ensure free flow at the high rates needed for lime-burning is about 0.7 bar (gauge), but the data given in the appendices is for typical average samples at atmospheric pressure (760mm Hg) and at 15°C.

Although natural gas cannot be liquified at normal temperatures it may sometimes be economically possible to cool it to a very low temperature (-113°C) and transport it in bulk as a liquid at this temperature.

Like LPGs it is a clean fuel, that is it contains never more than a trace of sulphur. It is lighter than air and is not poisonous and so it is safer to use than most gases. Where it is available it is well suited to use in lime-kilns. The design of burners is a matter for a specialist, but once the correct design has been found and an installation has been properly commissioned natural gas is a trouble-free fuel. Properties of a typical sample are included in the table in Appendix 12.

Producer gas

Producer gas is a fuel which is made in a 'gas producer' by the partial combustion of coal or oil in a restricted air supply. Steam is sometimes also added. The gas is a mixture of useful fuels and inert components. The main fuel components are carbon monoxide (CO) and hydrogen (H_2) with just a little methane (CH_4). The bulk of the gas is inert nitrogen and there will usually be small amounts of carbon dioxide and oxygen. This is a 'lean' fuel — that is, it has a low calorific value which can be an advantage where an even heat distribution is important.

Theoretically a gas producer should be an effective way to use the cheaper bituminous coals for lime-burning, but there are difficulties and after wide use in the first half of the twentieth century gas producers gained a poor reputation. There are two main design difficulties: to deliver the gas at a high enough pressure to get a good penetration into the burden, and to avoid choking the flues and contaminating the lime with impurities.

Gas producers for coal fall into two broad classes: they may be built with or without gas cleaning and cooling systems. In the simpler class of producer the hot gases carry over tar and soot and when these are deposited in the ducts or flues they reduce the performance of the burners. When the equipment is used for continuous production this is a serious drawback as running maintenance is difficult. Ammonia may also be produced and this contaminates the lime. On the other hand, cleaning and cooling systems are expensive and have generally only been used on very large installations.

Coal is re-emerging as a very important fuel and there is widespread research into new ways of converting it into liquid and gaseous fuels. The fluidized bed technique could be used for the gasification of coal although there is no commercial experience for lime-burning.

Research is needed to develop a small-scale gas producer which can work reliably using the cheaper grades of coal. The operating pressure should be about 0.7bar and the equipment should have simple ash management mechanisms and should include a gas cleaning system which is both cheap and easy to maintain.

Although producers may be designed for a wide range of coals, the important properties to expect of a suitable coal are good calorific value for money and not too much ash. In practice the choice will propably be a bituminous coal with a volatile content in the range of 30% to 40%. The Sayler Chart shows that such coals would have a gross calorific value of between about 34 and 35.6 MJ per kg. Different coals will produce somewhat different gases but the variations are not likely to have a great effect on the operation of the lime-kiln.

The entire installation from the gas producer through the cleaning system and ductwork to the kiln itself must be designed as a whole to achieve good results. As some of the coal will be burned in the gas producer to provide the heat to gasify the remaining combustibles, the operating costs must be assessed on the total amount of coal fed to the producer in relation to the lime which it makes. Capital and maintenance costs will also be very significant.

Some of the chemical and physical properties of a typical produc-

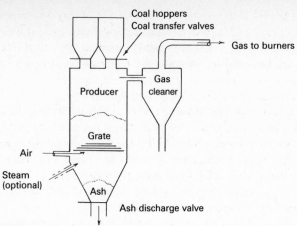

Scheme of gas producer using coal.

er gas are included in Appendix 12. The designer and maker of a gas producer will be able to give performance calculations and may be able to run trials using samples of the coal to be used.

Gasification of Oils
Oils can be used in gas producers in a similar way to coals. Oil gasifiers are sometimes called 'sub-stoichiometric combustors' meaning simply that the oxygen supplied to the oil is not sufficient for complete combustion. This has been found to be a useful method of burning oil in smaller lime-kilns. The diagram below shows a small, simple and cheap oil gasifier. Its simplicity makes it easy to manage and use. The design allows kiln exhaust gases to be

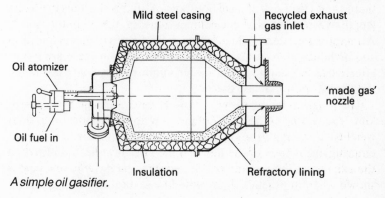

A simple oil gasifier.

recycled for flame attenuation and for temperature contol in the kiln. Appendix 12 includes some of the properties of the gas made during trials with a 3500secs heavy residual fuel oil. Other trials were run using lighter oils and the results were very similar.

How oils and gases are used in lime-burning

To get good value out of these rather expensive fuels it is necessary to invest in a more elaborate kiln than is needed for solid fuels. In most modern oil and gas fired kilns the fuel is introduced through a ring of ports set around the kiln. At least one kiln has a central burner and a few older kilns have both radial ports and a centre burner. As the burners are usually visible outside the kiln these kilns are often classed as 'externally fired'.

All of the oils and gases described above will ignite at temperatures between 460°C and 700°C and so once a kiln is operating at normal temperatures the ignition is self sustaining. A special arrangement must be used to start up the kiln when it is cold. Some kilns have permanent pilot burners whilst a temporary torch or flare must be used in others.

Lime-making starts when the surface of the carbonate reaches a temperature of around 900°C. Commercial lime-burners work at temperatures of up to 1250°C, to obtain a good heat penetration into the lumps, but above this the lime will certainly be spoiled. The most reactive limes are produced at the lowest temperatures.

If hydrocarbon oils and gases are burned with just the amount of air needed for full combustion then the flame temperatures are around 2000°C, which is far too high for lime-burning. Excess air could be introduced to lower the flame temperature but even then there would be a risk of local hot spots. Also this shortens the flame length which makes the calcining zone in the kiln too short. Attempts have been made to lengthen the flames by steam injection and by burning an emulsion of oil and water, but these are not as effective as the major alternative technique of recycling a portion of the almost inert exhaust gases.

The diagram below shows a flow of gases through a simplified kiln. The widths of the bands of gas flow are in proportion to their weights. For simplification *all* of the combustion air is shown as entering at the base of the kiln and the fans have been omitted on the exhaust at the top and on the recycling path. Only one port is shown when in practice there would be several. The supply of air

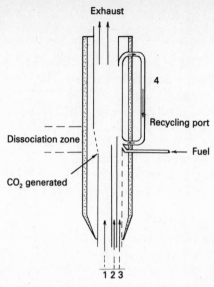

Diagram of gas flows through a kiln.

and the recycling are carefully balanced to get the best flame characteristics and to make best use of the heat. A simple calculation shows that about 3.7kg of combustion gases at 1250°C would be needed to convert the carbonate to make 1kg of quicklime. The proportions are arranged to achieve this by burning the correct amount of fuel for the planned output of the kiln. The fuel enters the kiln at the base of the dissociation zone and just below the recycling port on this diagram. The exact amount of air theoretically needed to burn this fuel is represented by the stream labelled '1' on the gas flow diagram. This is the stoichiometric weight. To be sure of complete combustion a little extra air must be allowed and this would be about 25% of the stoichiometric weight. It is represented here by stream 2. About a further 25% of the basic weight would also be needed to lower the temperature of the flame. This could be achieved by introducing fresh air at the base as shown in stream 3, but the more effective way is to use part of the inert exhaust gases as shown at stream 4 on the recycling path. These would be taken from a high level in the kiln at a temperature of about 350°C. After cooling in the duct and fan they would return to the kiln at about 250°C.

As air passes up from the base of the kiln in streams 1 and 2 it will be preheated by the hot quicklime before it reaches the dissociation zone.

At the dissociation zone carbon dioxide gas is released from the carbonate and this increases the gas flow. This is shown, to scale, by the widening of the stream.

To summarize, when gas or oil are used as fuels for lime-burning the gas flows in the kiln must be very carefully balanced to obtain the best results. This involves ductwork and pumps which raise the kilns above the lowest levels of technology. Oils may be converted into producer gas in sub-stoichiometric combustors as a way to avoid the risk of hot spots and to get a thorough combustion.

Possible alternative fuels

All of the fuels described so far in this chapter are relatively high energy fuels, that is they have a high calorific value and substantial denisty. With the exception of wood they were all fossil fuels of which the reserves must be limited.

The general search for alternatives to fossil fuels follows two paths. Nuclear energy is essentially a high technology and most unlikely to generate power cheaply enough for lime-burning. There is also the exploitation of renewable energy sources which all depend in some way on energy from the sun. This can be directly used for low intensity solar heating (with temperatures far too low for lime-burning), with photovoltaic cells to create electricity and with arrays of mirrors to concentrate the sun's rays. Indirect methods include the harnessing of wind and wave power. None of these seem at all promising for lime-burning in the near future.

What does seem promising is the use of biomass, that is the bulk material from plants, either as waste products or as plants cultivated for this purpose. Plants are formed by photosynthesis in which the sun's energy is captured. There are several methods by which the energy may be released and these will be described in the following sections.

Another source of alternative energy is low grade fossil fuel which is not of sufficiently high energy value to be too expensive. There is a long tradition of burning lime (with great difficulty) using peat and from time to time the deposits of shale oils have been used as an energy source for other industries. Both of these materials can be handled quite easily by modern combustion techniques, but they are very bulky.

Fuel prices can increase suddenly and the lime-burner must always remain aware of the possible alternative fuels.

Methods of using biomass energy
The cellulose bulk of plants has an energy value which is often high enough to make a useful fuel. There are several ways in which this can be used.

1   Incineration, that is straightforward burning.
2   Pyrolitic incineration for gas production.
3   Methane gas production by anaerobic digestion.
4   Briquetting.

The first technique is that used in the dome kiln described in Section 6.1. It is not very satisfactory for lime-burning and is remarkably hard work. The second is a more sophisticated approach. The biomass is burned in a deep bed in a furnace with a reduced air supply in much the same way that coal is converted to gas in a producer. There is research into this but no firm conclusions are available.

There has been very widespread research and development of methane gas production techniques and the technology must offer scope for use in lime-burning in the very near future. Methane was considered in the context of natural gas in Section 4.4 and it is a high calorific value fuel which is safe and easy to use. By the standards of other chemical works a methane production plant is simple, safe and easy to operate. Production is possible from a very small scale upwards. Even large plants present no serious design problems.

The molecular structure of cellulose can be broken down by a family of bacteria known as methanobacter. They convert it into methane and carbon dioxide gases and they leave behind a residual sludge.

At the centre of the operation are the digestors — enclosed tanks each divided into subsections. The gases are produced here from a shredded feedstock and as they are taken from the tanks they are pumped to a higher pressure for separation in a scrubber. The scrubber can remove as much of the carbon dioxide as is appropriate for the eventual use. For lime-burning much of it may be retained to make a 'lean' fuel. The methane is then stored in tanks for transmission by pipe to the lime-kiln or other end use.

It is likely that 'energy farms' will grow and harvest the crops as well as producing the methane. There the sludge can be reused as a fertilizer and the movement of the bulk materials can be kept to a minimum.

The fourth method of using biomass is by briquetting. This is the compression of the material to form hard cohesive lumps which can be used like wood or coal in a mixed feed lime-kiln. The raw material is first crushed and then dried before it is compressed by a piston in a cylinder. The density of the briquettes is very much greater than the bulk density of the feedstock. In some cases it may be as much as twelve times more. This not only makes combustion simpler but means that storage and transport of briquettes is much cheaper than handling the bulk materials. Excellent results were obtained in Mauritius when briquetted bagasse (sugar cane waste) was used for lime-burning.

Source of biomass

Most forms of agriculture and forestry produce waste materials which are usually just burned for convenient destruction. Some farmers now have special boilers to burn baled straw for their own heating systems and there are examples where whole islands have run their transport and industries on power from burning sugar cane waste.

The calorific value and cellulose content of these waste materials varies considerably and the low grade materials may always be burned off in this way. The better materials can be a source of fuel for the conversion methods described above, though bulk handling will be a problem.

The use of energy farms was suggested in connection with methane production. These would be farms growing those plants most suitable for degradation by the methanobacters. Biomass varies greatly in cellulose content and hence in potential gas yield. The list which follows shows some of the more important plants and includes a crop suitable for most regions where a gas farm might be possible:

Bahia grasses
Bermuda grasses
Giant brown kelp

Savannah grasses
Sugar cane and related grasses
Swamp and marsh vegetation
Sorghum
Water plants from river estuaries.

It must be emphasized that only a crop suitable for the local environment should be grown. Care should also be taken not to overwhelm the local natural wild plant and animal life which is particularly vulnerable in the wetlands.

Low grade fossil fuels: peats and oil shales
Peats and lignites are the lowest ranking members of the coal family. They were described briefly in Section 4.3 but they cannot be burned in the same way as other coals as their volatile content is so high.

Attempts have sometimes been made to use these fuels in gas producers but on balance it has usually been easier to find other, dearer, fuels which are simpler to manage.

In some circumstances peat is the only available fuel. In many countries there was a lingering tradition of lime-burning with peat using flare kilns but this very simple technology is wasteful of fuel resources. There are two modern techniques which can help. The first is briquetting which was described in the previous section. The loose material can be compacted into firm and dense lumps.

The other technique is fluidized bed combustion. This was developed for coal but is particularly suitable for low grade materials. A bed of ash or other incombustible material is violently agitated by an upward current of air forced through perforations at the base of a container. The air current is adjusted to make the material take on the appearance of a boiling liquid. Into this fluidized bed the fuel is continuously introduced whilst a corresponding amount of ash is withdrawn from the base of the bed. The whole bed acts as a fire into which boiler tubes may be set. Hot flue gases are taken from the top of the container and could be introduced to the kiln. When coal is burned, and this could include the low grade coals, it is usual to keep the bed temperature as low as 900°C to avoid problems of sintering. This means that the exhaust gas temperature is just too low for lime-making. For this the starved air technique has been used in New Zealand and the equipment is used as a gas producer to make the usual mixture of carbon monoxide and the other components of producer gas for final combustion in the kiln.

The fluidized bed technique has successfully burned peats, coals with very high ash content, and oil shales. In all of these cases the trials were showing 98% combustion efficiency when the bed temperatures was kept to 800°C. At such low temperatures the gases from complete combustion would not be suitable for lime-burning (unless perhaps some further fuel were introduced in the kiln?). The partial gasification techniques may well be possible with these materials and only a little further development is needed to provide a very useful tool for the lime-burner.

CHAPTER 5

Kilns

This chapter shows the extraordinarily wide range of kilns which are used to do very similar tasks. It is possible to build kilns by hand with available materials or to invest many hundreds of thousands of US dollars on high technology tools. There is generally an increase in fuel efficiency and sometimes a fall in labour requirements with the increased sophistication. For each project the appropriate kiln must be chosen and this must take account of several factors:

Quality of lime required
Available capital
Available expertise
Available labour
Available fuels
Availability of spare parts
Construction and maintenance skills
Total demand and pattern of demand
Fuel costs
Other running costs.

A selection of kilns is described in an approximate order of technical sophistication. The earliest examples are very wasteful of fuel and the later examples are extremely expensive. In the middle ground are the types of kiln which will generally be appropriate for small scale projects, but there are lessons to be learned from the more extreme examples.

5.1 Very simple kilns

Uncovered heaps
The simplest and most extravagant method of burning lime uses no kiln. Alternate layers of wood and limestone are heaped up to a height of about 3 metres with an extra hump of stone at the centre. There is no control whilst the fire burns through the pile.

An uncovered heap with alternative layers of wood and limestone.

If wood is used without coal there should be very little overburnt lime even at the centre, but around the edge there is certain to be a substantial amount of underburnt lime.

More than a tonne of wood must be used to produce each tonne of lime, but as there is so little capital outlay this very inefficient method has often been used.

Covered heaps

An improvement on the uncovered heap is to build the alternate layers of wood and stone as before, but to cover the whole heap with

A heap covered with clay daub.

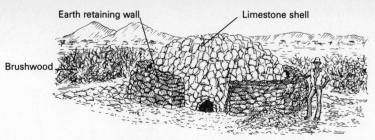

Earth retaining wall   Limestone shell
Brushwood

The Cape Verde Islands dome kilns.

a daub of clay leaving air inlet holes around the base and an outlet at
the top. This helps to conserve some of the heat and gives a little
scope for controlling the speed at which the wood burns, but the
clay skin soon fails through shrinkage cracks as it dries out. This
improvement shows how engineering can begin with very simple
materials.

Dome kilns

A further method of lime-burning without a permanent kiln is to
build a hollow dome out of lumps of limestone and to maintain a fire
inside the dome until the stone is converted to lime. This requires
great skill from the lime-burner, both to build the dome and to
judge the completion of the firing. The authors know of only one
example of this method and there, on one of the Cape Verde
Islands, the lime-burner moves from site to site following the
availability of salt marsh scrub for his fuel.

He builds his domes freehand within a 4 or 5 metre diameter
retaining wall and using no temporary supports. He fires the kiln for
a little over 24 hours using only the dried scrub. Up to 600m^3 of the
scrub are used for each firing. After firing the dome collapses into a
shallow pit, and the quicklime is slaked in this pit. The whole cycle
of gathering materials, building the dome, firing slaking and bagging lasts about one month.

With such unconventional fuel it is impossible to make any
judgement about the fuel efficiency. This design shows how ingenu-
ity can make use of the most unpromising fuels.

Pit and beach kilns

In some situations lime can be burnt in pits dug into the ground. The
conditions are often found in unsorted beach sand, volcanic ash and

A pit kiln before loading.

tuff. In many tropical countries coral is burnt on the sea shore in pits dug into the sand above the high water mark.

The pits are typically around 2.75m long × 2.50m wide and 1.5m to 2m deep. In the middle of one side a trench my be cut leading down to a doorway into the pit. This would be reinforced with two stone cheeks and a further stone would form the door itself. The door would be set in place when the fire is established.

The pit is filled with a layer of kindling and then with alternate layers of timber and carbonate. This is cut into well graded lumps perhaps 10cm to 15cm across. The charge might be heaped up well above the ground level as a mound. When this method was used in England in the nineteenth century the lime-burners would build up low walls of turf and cover the heap with further turves.

The fire is lit and will burn for 24 hours or more. If no trench is used the heap is formed with perhaps six lighting ports around the edge. These provide much of the draught for the fire. If the flames

Cross-section and view.

burn through the upper surface during the first day, further limestone or coral is added to cover them.

As the fire burns, all the materials settles into the pit where the heat is retained for several days. When it has cooled the lumps are carefully picked out. The unburnt stone can be sorted from the good quicklime by its heavier feel.

The fuel is often a mixture of fast and slow burning woods. Coal or peat may be added. The fuel consumption is high, perhaps the equivalent of 15m³ of palm tree logs would be needed to produce 5m³ of quicklime. The cost of the kilns is very low. This simple kiln can use coral and shell which can be difficult to burn in the more efficient kilns.

5.2 Simple kilns

Flare kilns
Throughout the world there are simple lime-kilns used for batch production. The shape, materials and size of these kilns varies so much that it is hard to believe that they all serve the same purpose. Their common feature is that limestone is placed inside a masonry kiln over one or more grates on which fuel is burnt. When the charging is complete the lime-burner lights the fire and keeps refuelling it until he judges that all the limestone has been converted to lime.

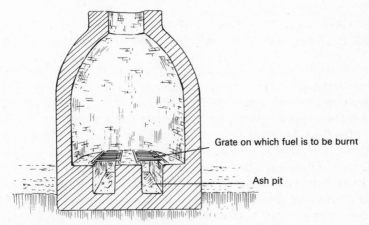

Grate on which fuel is to be burnt

Ash pit

Cross-section of a flare kiln with two grates.

The efficiency of these kilns depends very much on the skill of the lime-burner, but there are severe limitations which he cannot overcome. The whole fabric of the kiln must be heated up from cold on every firing. At the end of each firing this heat and the heat stored in the hot fresh lime will be lost when the charge cools down. Fuel consumption might be about half a tonne of coal for each tonne of lime.

Good quality lime can be produced in flare kilns but there are difficulties. The heat is likely to be most intense around the grate so there is a danger of overburning the lime near to it. If the fire is maintained for too long the overburning could extend to much of the charge. On the other hand if the fire is stopped too early much of the stone will be underburned.

Pot kilns

The terms used to describe flare kilns are not well defined but one type is often called a pot kiln. In these kilns the masonry structure is usually between 3 and 8 metres high. The grate (there may be more than one) is about 500mm wide running right across the kiln. In front of the grate is a 'draw hole' about 1.2 metres high. The internal shape is bellied with a maximum diameter of about half the height and narrowing to an opening at the top. This top opening would be about one third of the maximum diameter.

Many other shapes are found ranging from double truncated cones to simple cylinders and square or octagonal prisms. The shape is not very important.

The lime-burner places an initial charge of wood and coke or coal on the grate and builds a rough arch or dome of large stones over it without any mortar.

Above this arch he places progressively smaller stones with the smallest pieces at the top. His object is to leave a network of voids through which the flames can heat the kiln evenly. Slabs of stone are sometimes set across the top to regulate the draught.

The fire is lit and maintained until the lime-burner judges that all the stones has been converted to lime. This is usually about 48 hours after the fire has reached its greatest heat or 72 hours after lighting. There is no certain way of knowing when the stone has all been converted but there are two important clues. For many materials the lime will be much softer than the original limestone or chalk. A poker can be forced into the quicklime where it might be impossible

A pot kiln.

Draw hole

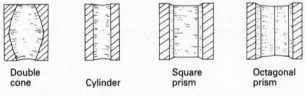

| Double cone | Cylinder | Square prism | Octagonal prism |

Common veritical sections of pot kilns.

The rough dome.

Stones to control draught.

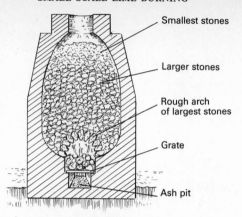

Cross-section of a loaded pot kiln.

to push it through the raw materials. The other clue is the shrinkage which can be seen as the conversion progresses.

When the burning is stopped the kiln is allowed to cool down until the materials can be handled. The ash is removed and then the lime is withdrawn through the draw hole.

The possible use of batch production methods is not usually explored in industrialized countries, but they can respond more readily to special needs where capital is low or where demand is not constant.

Hillside kilns

Pot kilns are often built against a hillside to provide easy access to the top for loading. Soil or broken limestone may be packed around the kiln to increase the insulation and to provide a working platform. This also tends to increase the thermal capacity partly offsetting the advantages of insulation as a greater mass must be heated up and allowed to cool at every firing. A retaining wall is needed to hold back the loose fill and a tunnel must be formed to give access to the draw hole.

Very simple hillside kilns are sometimes found. These are just a gash in the hillside with a brick or stone lining and a retaining wall at the front. The front wall might be built up as the charge of limestone is carefully placed.

A hillside kiln.

Climbing kilns

Although simple in construction, climbing kilns demonstrate an important advantage which is normally only found in continuous production kilns. When the burning is complete in a flare kiln the lime and the kiln are both very hot. This stored heat is usually wasted on every firing but in climbing kilns a way was found to recover part of it. A sequence of small kilns is built up the slope of·a hillside. Each kiln is connected to the next by a tunnel which acts as a chimney to take the exhaust gases from one kiln up to the next. These tunnels are formed of clay daub on a light framework of sticks. The kilns are small dome shaped chambers perhaps 2 metres in diameter.

All the chambers are filled with fuel and limestone and the lime-burner lights the fire in the lowest chamber. As this burns through, the exhaust gases pass up the tunnel heating the next chamber and even the ones beyond.

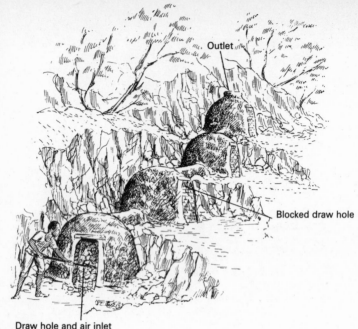

Climbing kiln.

After an appropriate delay the fire is lit in the second chamber. Much of the air to support this is drawn through the tunnel from the lowest chamber, although a proportion of fresh air may also be introduced. The hot gases from this chamber pass up through the tunnel to the higher chambers. When the fire in the lowest chamber has burnt out, air is still drawn through it to feed the chambers above. This flow of air takes much of the stored heat out of the fresh lime and helps to heat up the limestone in the higher chambers.

The process continues as the higher chambers are each fired in sequence. The work of clearing out the lime and recharging the lower chambers can begin whilst the upper chambers are still being fired.

There are no figures for the fuel consumption of these kilns but they should be very much more efficient than a normal flare kiln.

Intermittently fired mixed feed kilns
Some hillside kilns are built as a short cylindrical shaft, perhaps 6

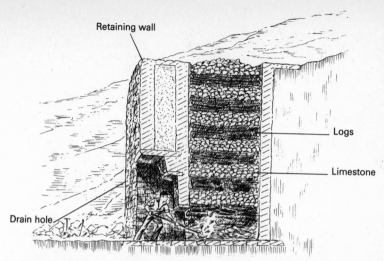

Intermittently fired mixed feed kiln.

metres high by 2 metres diameter with a draw hole at the base but without a grate. A layer of brushwood is laid at the base on a bed of peat or coke and the shaft is filled with alternate layers of limestone and fuel. This is lit and allowed to burn through without further attention. When the charge has cooled the lime is drawn out.

The method uses considerably more fuel than a well managed pot kiln but does not require the skills of an experienced lime-burner. These kilns are often used on farms where lime is required only once or twice a year and where a low grade product with ash, underburnt and overburnt lime is acceptable.

Running kilns or draw kilns
Kilns very similar to those just described are also used for continuous production. They are called running kilns or draw kilns because of the way they are used.

The kiln is charged in much the same way as for the intermittent use already described. When the lime-burner judges that the stone at the bottom has been converted to lime he draws out some of the material through the 'eye'. The level of the materials in the kiln is maintained by adding further alternate layers of limestone and fuel. The whole bulk should settle down, but sometimes the material forms an arch across the kiln and is said to hang. The lime-burner

Charging

Drawing quicklime

A similar kiln used as a running kiln.

breaks this arch by poking the stone through the poke holes which are always provided in draw kilns. The kilns may be kept burning for up to a year before they are run down for repairs.

Fuel efficienty would be better than for a comparable flare kiln — each tonne of coal might yield 2.5 to 3 tonnes of lime. This, again, depends very much on the skill and judgement of the lime-burner.

The reason for the increased efficiency is that there is some scope for the fire in the middle of the kiln to preheat the freshly loaded limestone and fuel at the top. Also the air drawn through the converted lime and ash at the bottom of the kiln is warmed up before it reaches the fire in the middle. Unfortunately if the height is less than about 6 metres there is not enough scope to make the most of this useful heat transfer as the intense fire will be quite near to the top of the kiln. Indeed it is only the skill of the lime-burner which prevents it from burning through the top stones.

5.3 Improved kilns

Improved simple shaft kilns with mixed feed or other firing
The draw kiln was the most successful of the simple kilns for commercial use. It has been developed in various ways throughout

the twentieth century to suit different levels of production, different raw materials and fuels. The most obvious division in this family of kilns is between those fired by solid fuel mixed with the limestone charge — known as mixed feed shaft kilns and those in which the fuel is burned separately. Separate firing prevents contamination of the lime with ash but leads to other problems, especially the risk of overburning at hot spots in the kilns. Several shaft kilns are available from manufacturers. These are usually cased in steel and of large capacity with outputs in the range 20 to 250 tonnes per day.

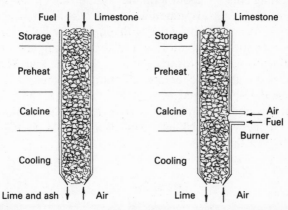

Vertical sections through mixed feed and separately fired vertical shaft kilns.

The principle underlying the design of shaft kilns is the recovery of heat by the upward passage of air through the kiln. The height of the shaft is considered as containing four approximately equal zones. There are no physical boundaries between these zones and it is only by the careful operation of the kiln that their identities are retained. The charge enters the kiln through a suitable device into a storage zone at the top. The purpose of this storage zone is to ensure that the preheating zone below it is always full. It will probably be rather less than a quarter of the height of the kiln. Automated kilns may have a radioactive beam cutting across this zone from an emmitter on one side to a sensor on the other. When the stone falls below the level of the beam a further charge of raw material is introduced.

In the preheating zone the hot gases from the lower parts of the

kiln dry out the charge and heat it up to near the dissociation
temperature. In a mixed feed kiln the fuel should not be hot enough
to ignite in this zone. By the time the kiln gases pass up into the
storage zone they should have cooled to a temperature as low as
300°C.

The charge falls gently through the preheating zone into the
calcining zone where the heat is delivered from the fuel. In a mixed
feed kiln the lime-burner aims to restrict the combustion to this
zone. In separately fired kilns the fuel ports will be at the bottom of
this calcining zone — about a quarter of the way up the kiln. The
lower part of this zone is sometimes termed the finishing zone. The
calcining zone must be provided with ample poke holes for inspection or instrumental monitoring and for breaking any arches or
'scaffolding' which may form in the material in the kiln. Some kilns
are designed with the deliberate intention of forming such arches.
When the lime-burner judges from the colour that the material in
the arch is completely fired, or in danger of being overburned, he
breaks it to allow the quicklime to fall down into the cooling zone at
the bottom of the kiln.

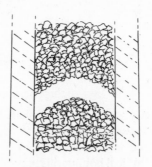

'Arching' of the charge in a shaft kiln.

In the cooling zone the hot lime is cooled by the incoming air
until, ideally, it is only hand hot when it is withdrawn from the
bottom of the kiln through the discharging device. This flow of air
through the kiln has to be regulated carefully. In some patent kilns
there are ducts and fans to take a portion of the kiln gases from the
cooling zone back to the dissociation zone.

The possibilities for developing this simple principle seem end-
less. A form designed and manufactured by Wests Pyro Limited is

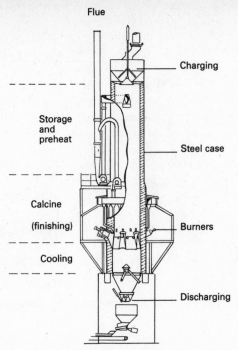

Flue

Charging

Storage
and
preheat

Steel case

Calcine

(finishing)

Burners

Cooling

Discharging

A separately fired shaft kiln designed by Wests Pyro Limited.

illustrated as an example. A research programme in India during
the 1960s produced a shaft kiln design which is of much more
interest to the small scale lime-burner. This is the Khadi and Village
Industries Commission kiln which has an output of 1.5 to 2.0 tonnes
per day. The significant features of the design are shown on the
illustration. The greatest single improvement over the simple running kilns described in Section 6.2 is the increased height which
allows adequate cooling and preheating.

In the first half of the twentieth century one of the most successful
improved shaft kilns in England was the Brockham kiln. Kilns of
this simple type are still used for burning chalk with coal. The chalk
and low grade coal are charged as mixed feed through a door in the
brick chimney. The shape swells out suddenly at the calcining zone
and this allows direct inspection through sloping tubes around the
kiln. Small quantities of higher grade coal can be introduced
through the inspection chutes to give fine control of the firing. This

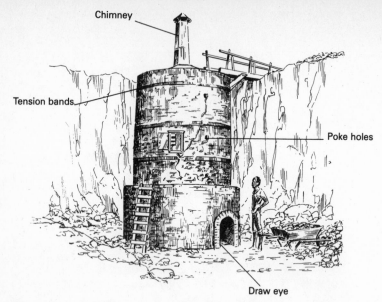

The Khadi and Village Industries Commission (KVIC) improved shaft kiln.

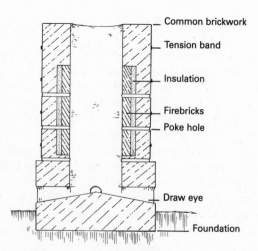

Section through KVIC improved shaft kiln.

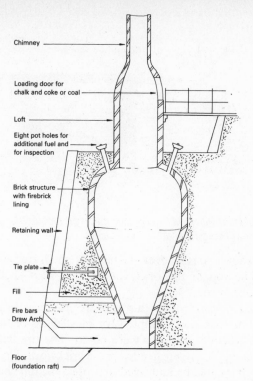

Chimney

Loading door for chalk and coke or coal

Loft

Eight pot holes for additional fuel and for inspection

Brick structure with firebrick lining

Retaining wall

Tie plate

Fill

Fire bars
Draw Arch

Floor
(foundation raft)

A late form of the Brockham kiln.

operation is labour intensive but effective. Very good quality lime can be produced.

Improved flare kilns

The heat recovery mechanism in a shaft kiln gives it an inherent commercial advantage over a flare kiln. For this reason little progress has been made in the development of flare kilns, but there are still many situations where flare kilns would be appropriate. This is particularly so where only a small output is needed and practical economics will often point towards a small output.

New flare kiln designs have been experimental and without further development they may not offer significant improvements over a well managed traditional flare kiln. The kiln used by the Transport and Road Research Laboratory and the Building and Road Research Institute, Kumasi, in Ghana in 1972 is illustrated. It

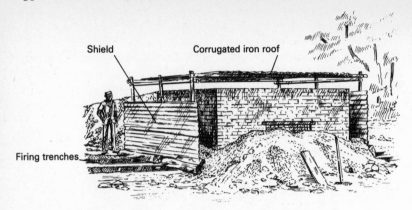

A very simple flare kiln for a road construction project in Ghana (Transport and Road Research Laboratory).

is low and square. It is set over two hearths which are trenches in the ground. The top of the brickwork is open but sheltered by a corrugated iron roof.

The main advantages of this kiln are its low cost and simple construction. It is easily adapted for most types of fuel. Several minor construction problems were noted. The fire bars distorted and movement joints were needed in the brickwork. The ends of the firing trenches could be improved to protect the lime burners from the heat, to regulate the draught and to allow greater variation in the length of logs. The roof sheeting failed and this could be remedied by improving the roof or by containing some of the heat by covering the top of the charge with a layer of open jointed bricks.

Fuel consumption of around one tonne of good logs for one tonne

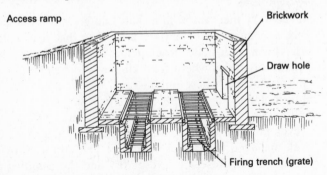

Section through the TRRL and BRRI flare kilns in Ghana.

of quicklime produced was expected of this design. The rectangular shape gives rise to cold areas within the kiln, especially in the corners and at the bottom away from the firing trenches.

A later flare kiln was designed by Halcrow Caribbean Ltd. for use in Belize. This avoided the corners by choosing an inverted truncated conical cross-section which was harder to construct. But this traditional shape has the disadvantage of a large surface area which allows excessive heat losses through the open top.

The fundamental disadvantages of the flare kiln are that the whole mass of the kiln must be heated and cooled on every firing and that there is no recovery of the heat available in the hot lime when it is produced. Insulation is essential to minimize the heat loss from the fabric and every effort should be made to keep the thermal capacity of the kiln as low as possible. Heat recovery from the hot lime might be achieved by developing the principle shown in the climbing kilns described on pages 77–8.

Inclined chimney kilns

An inclined chimney kiln is a development of the flare kiln. The type is traditional for pottery in Korea but a demonstration kiln was developed by VITA and others in Honduras in 1976.

An arch or vault of good quality bricks was built up the 12° slope of a hill. It had a catenary section (that is, the shape in which a chain would hang naturally, but formed the other way up as an arch) about 1.8m high and 1.8m wide. The length was 20m. At the upper end the vault was connected to a 5m high chimney. The two ends were open for loading and unloading and bricked up during the fire period — leaving sufficient openings for entry of air at the lower end. Along the length of the kiln there are firing holes where fuel could be added during the burning. These were normally sealed. There was a thatched roof sheltering the kiln. Limestone was stacked in heaps between the firing holes. The first fire was lit at the lower end and this was followed by other fires, in sequence, higher up the kiln. In this way there was an element of preheating and heat recovery similar to that in a set of climbing kilns.

The kiln was found to have a fuel economy up to 40% better than that of the local Honduran pot kilns and the capital cost was about 20% higher than the equivalent capacity of pot kilns. On the other hand the inclined chimney kiln had a capacity four times greater than the pot kilns. The construction was demanding but required

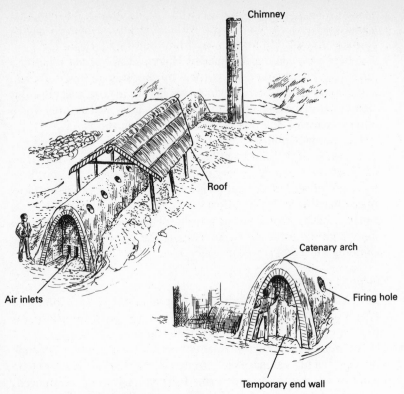

Inclined chimney kiln, Talanga, Honduras.

good masonry skills rather than high technology engineering.

Ring kilns
The Hoffman ring kiln was developed in the nineteenth century and used for lime, cement and brick manufacture. The kiln is still common in brickworks but is no longer in general use in lime or cement manufacture. An example reported in Greece seems to have been built originally for brick production. These are essentially large kilns but are described here because the underlying principle is so important.

Of the many variations developed, the one illustrated is a large brick structure with a ring shaped vault divided into twenty chambers each similar to the others. For convenience the ring may be distorted into any plan shape and was commonly rectangular. Each chamber is linked by arches to the two adjacent chambers and is accessible from the outside through a further arch. Smaller openings are provided for the addition of fuel through the top, for venting the flue gases through the ducts to a chimney and for introducing fresh air. The tunnel is about 60 metres in length and perhaps 3m high. Production is continuous and at any one time the uses of the chambers will be these:

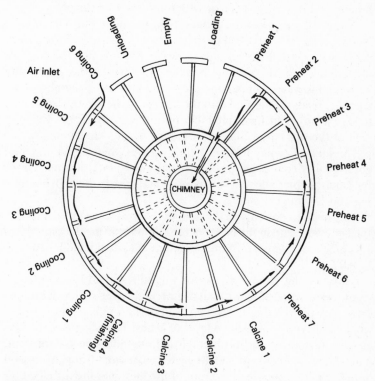

Diagram of chambers in a ring kiln.

One chamber is empty.

The next chamber is being stacked carefully by hand.

The next seven chambers are being preheated by hot gases from the fire. The first of these (and only the first) will be vented to the chimney.

The next four chambers are under fire with the lime-burner adding fuel through the traps.

The next six chambers are cooling and at the same time heating the air passing through the rest of the kiln. The fresh air will enter through the damper in the last of these.

The last chamber is being unloaded.

As each chamber is loaded, the arch between it and the empty chamber beyond is sealed with a temporary wall of paper. When the loading is complete the arch to the outside (the draw eye) is closed up completely with a door or with temporary brickwork.

When the lime-burner judges that firing in one chamber is complete he 'takes a fresh chamber' by opening the chimney flue damper in the recently stacked chamber and opening the fresh air inlet into the fifth cooling chamber. He breaks the paper wall between the new chamber and the other preheating chambers with a poker, feeds a little fuel into the hottest preheating chamber and feeds the adjacent firing chamber more intensely. The whole pattern of the operation thus moves along by one chamber.

The fuel efficiency for these kilns was very high. Each tonne of good quality coal could produce about 6.5 tonnes of lime. This is because the form allowed very good heat recovery through preheating and cooling chambers.

This applied not only to the heat in the fresh lime but to much of the heat stored up in the fabric of the kiln. However, the capital cost was high and a very substantial amount of skilled hand labour was needed.

Three of the variations on this ring kiln should be mentioned. Some kilns were built without the arched covering. A temporary cover was provided by laying two layers of firebricks over the charge and smothering these with a layer of ashes. This increased the heat losses and made the kiln vulnerable to rain, but as the kiln top could fall when the lime contracted it gave a more even firing (hot gasses otherwise tend to follow the roof of the kiln) and the form had the great advantage of allowing mechanical loading and unloading with a mobile grab crane.

A few kilns were built in suitable soil conditions by excavating the chambers below ground. This avoided much of the structural masonry, but left the lime vulnerable to damp penetration.

Where the market conditions demanded a pure lime without any contamination from ash, the kilns were built with a separate firing grate below or alongside each kiln chamber. This was the arrangement at the de Wit downdraught ring kiln built at Amberley in Sussex (England) in about 1905. This was a particularly expensive kiln which was never operated successfully by the staff at the lime-works who adapted it by building simple shaft kilns in four of the chambers. Similar kilns were operated competently in Belgium and the difficulty may have been the attempt to transfer a new technology without adequate instruction.

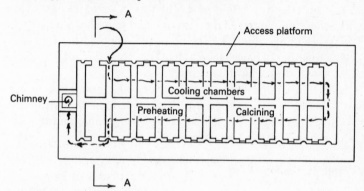

Simplified plan of a de Wit ring kiln.

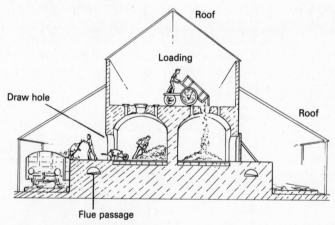

Section AA through a de Wit ring kiln.

To summarize, the ring kilns provide excellent fuel economy with simple technology although the scale of operations is always large with outputs around 45 tonnes per day. Labour and capital costs are very high.

Grouped shaft kilns

One further example may be given to demonstrate the overall principle of heat recovery which characterizes the best of these improved kilns. In a footnote to his textbook published in 1935 Searle mentioned that an ingenious means of increasing the thermal efficiency of chalk kilns had been devised by Priest's Furnaces Ltd. who had built four vertical mixed feed kilns close together. Each kiln had an effective height of 8.5 metres. The waste gases from each of these kilns entered the next one and so preheated the chalk in the succeeding kilns. The saving in fuel enabled the four kilns working together to have a lime to fuel (i.e. coal) ratio of 6:1. This is almost as good as a Hoffman ring kiln.

Chalk cannot normally be fired in a very high (and thus efficient) shaft kiln. In any case the construction of high shafts presents problems where only simple technology is available and this successful example is most encouraging.

5.4 High technology kilns

Rotary kilns

Rotary kilns are large long tubes which revolve slowly about an axis inclined at about 4° to the horizontal. The diameter is between 1.2m and 3.3m and the length is between 20m to 200m, usually nearer to the latter. The diameter-to-length ratio is about 1:30–40. The outputs are high — up to 1000 tonnes per day. The limestone is introduced at the top of the tube and fuel is blown in at the bottom.

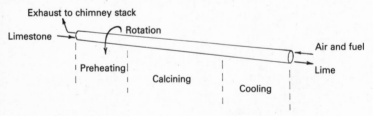

Diagram of a rotary kiln.

The tubes have a steel casing with a firebrick lining. The centre may be divided by radial webs to give, in effect, three parallel tubes within one steel casing. The casing revolves on trunnion bearings at a speed between 35 and 80 revolutions per hour.

The one outstanding advantage of this kiln is that it can handle small particles without the choking effect which they have in shaft kilns. For this reason a small rotary kiln will sometimes be found alongside more efficient shaft kilns in a large limeworks.

The thermal efficiency of the basic rotary kiln was appallingly low. It used perhaps 4000kcal/kg of lime, but this type of kiln is very widely used in the United States where fuel was relatively cheap. The efficiency can be greatly increased by a wide range of ancilliary equipment, but all of this increases the capital cost and maintenance demands.

The rotary kiln appears to be of no relevance to small-scale lime-burning.

Parallel flow shaft kilns

These complex and highly automated kilns have the highest fuel efficiency yet achieved. Two, or sometimes three, vertical shafts are worked together with downdraught firing. The multiple firing

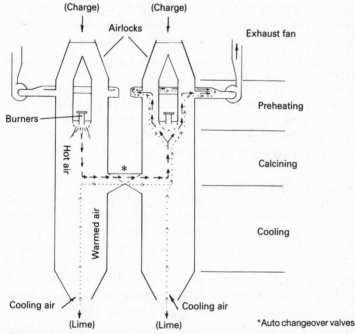

Cross-section of a parallel flow shaft kiln.

lances, burning gas or oil, play down near the top of one or other of the shafts and calcining takes place below the lances. The forced draught through the kilns is downwards through one shaft and then upwards through the second shaft. Flame lances play down this shaft for about twelve minutes. During a changeover period of about three minutes lime is withdrawn from this shaft and a fresh charge introduced at the top. The ducting is reversed and for the following twelve minutes the lances play down through the other kiln — the gas flow being completely reversed.

The controls are very precise so that little heat is wasted. The exhaust gases leave at only 120°C and the quicklime is drawn from the kiln at 65–90°C.

The calcining zones at the top of each shaft are insulated with 69cm of refractory linings.

The capital costs are very high and outputs of the kilns are high at between 100 and 600 tonnes per day. Fuel consumption is around 900kcal/kg of lime.

As the calcination temperature is kept very low — claimed to be in the range 950–1050°C — the lime is of the highest quality.

The present equipment is not suitable for the small scale lime-burner.

Double inclined vertical kiln

This is another successful high technology kiln. It is a vertical kiln with an extremely tortuous shaft. As the charge passes downwards it is tumbled this way and that so that small stones of 2cm or less can be calcined without choking the draught. The kiln is highly automated and yields a good quality lime at a good fuel efficiency.

Calcinating very fine particles

There have been several recent developments to enable very fine particles — even waste kiln dust — to be calcined. They involve quite complicated handling techniques and generally use fluidized beds in which the particles are kept in suspension by floating on air jets.

As the particles are so fine the kiln temperatures can be kept to the very minimum (870°C), so highly reactive lime can be produced. Fuel efficiencies are good but capital costs are high.

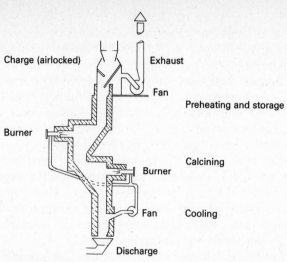

Charge (airlocked)   Exhaust

                            Fan

                            Preheating and storage

Burner

                            Calcining

                    Burner

                            Cooling
                    Fan

                    Discharge

Double inclined vertical kiln.

5.5 Possible lines of development

A wide variety of kilns have been described and most of these could
be developed further for effective small-scale lime-burning.

There has been very little recent development of flare kilns, yet
these provide the only effective method for very small-scale lime-
burning. There may be many locations where lime-burning could
begin on a very small scale and develop gradually to the level at
which existing continuous production kilns are viable.

A good flare kiln must have good insulation, good burning
arrangements and a low thermal capacity. This means that the
insulation must be placed close to the kiln lining. As there is no
friction, except from the contraction of the charge, it might be
possible to build the kiln walls themselves from lightweight insulating blocks.

The next stage up from the single batch kiln is likely to be the
addition of further batch kilns. These can be made to work with the
original kiln if they can be arranged as climbing kilns — with gases
passing from one kiln to the next to provide some preheating and
heat recovery on cooling. Better still, the kilns might be formed into
a ring on the principle of the Hoffman ring kiln. This ring could be
arranged so that further kilns or chambers could be added at a later
stage.

Flexibility is important. We saw that shaft kilns were used together to obtain from four short kilns the sort of benefits normally only obtained with much higher kilns. There has been some experimental work to prepare sintered PFA pellets for blockmaking using miniature shaft kilns, with diameters ranging from 100mm–1m. The experiments were not very successful but they did show that very small shaft kilns were possible. A large number of small shafts might be worked together for continuous production. This could avoid the problems of building and maintaining high structures. The small size of the individual shafts could mean that production could be expanded or contracted in small stages.

Much of the efficiency of the best modern kilns depends on accurate measurement and swift response of the controls. The use of microprocessors might make these advantages available to the small-scale lime-burner at an acceptable price.

Only just beyond the horizon of what now seems possible is the development of solar furnaces to burn lime from fine particles of carbonate. Solar furnaces are still extremely expensive and the technology of burning very fine particles is largely experimental.

CHAPTER 6

Practical Hints: Kiln Design and Operation

6.1 Rule of thumb to estimate the output of a kiln

Batch production

The output of a kiln for batch production (a flare kiln) will depend mainly on its internal volume. Each tonne of pure limestone could theoretically yield 0.56 tonnes of quicklime. As there is likely to be some wastage through underburning and overburning, perhaps only 90% of this is likely to be achieved. On the first trials with inexperienced operators the useful output may be far less than this. The bulk density will vary with each limestone and with the grading and packing of the pieces, but a likely figure is 1500kg/m³. On this basis the output for each batch will be:

Approximate quicklime output $= 90\% \times 0.56 \times$
for each batch $1500\text{kg/m}^3 \times$ usable
                                   internal volume

The time taken for each batch must be assessed to work out the monthly or yearly output. This must include the time to charge, fire, cool and discharge the kiln. If the lime-burner is working with very little help he should add the time to slake and package the quicklime and perhaps even to gather new stone and fuel before beginning the cycle again. All of this will depend on the way the labour is organized. The minimum time is likely to be one week and if the materials must be found the cycle may take a full month.

For a two week cycle and allowing the same 10% wastage 0.66m³ of usable internal volume will be needed for each tonne of output per month.

Shaft kilns

The output of a shaft kiln in continuous production depends largely on its cross-sectional area. The rule of thumb is that for each square metre of the cross-sectional area of the shaft the output should be about 2.5 tonnes per day. Restating this in the form given above, about 130cm² of cross-sectional area is needed for each tonne of quicklime output per month.

Even a well-operated kiln will need to be closed down from time to time for maintenance and repairs and between three and six weeks production may be lost in this way every year. In some climates the weather conditions might make production impossible at certain times of the year. A suitable allowance should be made when predicting the design capacity.

6.2 The height of shaft kilns

An efficient shaft kiln needs a good height to allow the quicklime to be cooled and the fresh charge to be effectively preheated. At all times the lime-burner must maintain four bands in the kiln to allow space for the separate activities of drying and storage, preheating, calcining and cooling. If the kiln is unattended at night these bands will drift upwards. When lime is drawn the bands drop down. If an adequate height is not available the fire may rise so high in the shaft that there is no preheating, or fall so low that there is inadequate

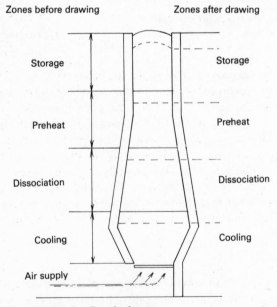

Four bands maintained as lime is drawn.

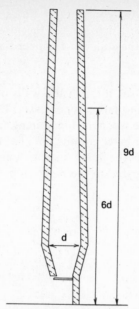

The height of a shaft kiln should be at least six times the diameter and preferably nine times.

cooling. Heat would then be wasted by drawing lime at too high a temperature.

It is not just the height of the kiln that is important, but also the ratio of the height to the diameter. A shaft should be at least six times higher than its diameter and preferably nearer nine times for greatest efficiency.

Thus a small kiln with an internal shaft diameter of 1.5m should have an internal shaft height between 9m and 13.5m. These heights cannot be used for soft materials which would crush under the pressure nor for fine grades where the gas flow would be too restricted. Special arrangements have to be made for calcining these. Shaft heights for soft chalk are often restricted to between 3m and 5m.

6.3 Foundations

Shaft kilns are necessarily tall structures which must have good foundations to ensure their stability. The foundations must be able to withstand the weight of the fully charged kiln without excessive deformation. On the quarry bed there should be little problem, but if the kiln is built on a sloping site the pressure must be evenly distributed through the soil. An engineer's advice is needed.

6.4 Insulation

Any heat which passes through the walls of a kiln is completely wasted so effective insulation is necessary. Designs often include a layer of insulation between the inner lining and the outer structural shell of the kiln. For batch production kilns this layer should be as close as possible to the inner face to minimize the volume of masonry which is heated to the high temperature at each firing.

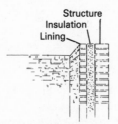

Batch kiln — insulation close to the inner face.

In shaft kilns the abrasion wears out the lining and so there must be rather more masonry inside the insulation layer. This is quite satisfactory as the mass of masonry need only be heated up once in every period of continuous firing. Special care must be taken in insulating the calcining zone in shaft kilns as the highest heat losses occur where the temperatures are highest.

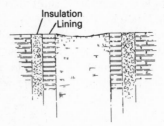

Shaft kiln — more masonry within the insulation.

The materials available for insulation vary from year to year but nearly all depend on their ability to trap air. Some insulation could be given to a kiln by simply leaving an air gap in the wall thickness, bridged by occasional bonding bricks to retain the structural integrity. This can be greatly increased by filling the gap with an effective insulating material. Any lightweight material which can withstand the temperature will be effective. Pozzolana or burnt paddy husk would be good, or perhaps diatomite, rockwool or the sintered slags which are sometimes used for lightweight concrete. A fine material may be able to flow into small cracks in the kiln to prevent loss of the hot gases but larger cracks may drain the insulating material into the shaft.

Where a kiln is to be built into a hillside the filling between the brickwork and the excavation should include a good insulating layer. Reasonably well-graded aggregate will also help the insulation and still give a firm filling. If the aggregate is kept clean without powder or fine particles, the airspaces can provide insulation.

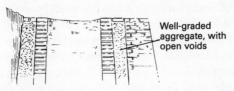

Well-graded aggregate, with open voids

Insulating a hillside kiln.

6.5 Structural masonry

The structural masonry of a kiln can be of brick or stone and should be formed by skilled masons who know how to bond the units and to form simple arches and corbels.

A good general standard must be maintained. The vertical joints should be staggered from course to course. Vertical and bed joints should be completely filled with mortar. Natural stone should be laid on its natural bed — that is with the formation layers or laminations horizontal. The kiln structure will experience a very much wider range of temperatures than any normal building. When heated the structure will expand and the hottest parts will expand the most. This means that a stiff structure — such as one with a cement mortar — would crack badly. A mortar with some give in it is needed. Pure lime and sand mortar is ideal. Small construction units, that is bricks rather than blocks, provide more mortar joints with the possibility of forming numerous harmless hairline cracks rather than fewer larger cracks. The quality of the bricks is not important provided they can withstand the normal weather conditions on the site. A well-burnt brick should be used for the top courses which would suffer most from weather.

Natural stone laid on its natural bed.

6.6 Linings

The choice of lining for a batch kiln is not likely to be critical as there is little abrasion between the charge and the lining. If bricks are used they should be fired at a temperature higher than the kiln temperature. Limestone was often used to build kilns but obviously this tends to calcine and waste away. Sandstone or granite would be more suitable than limestone but are unlikely to be available locally.

In shaft kilns the linings are very important as the continual

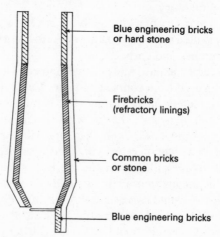

Blue engineering bricks or hard stone

Firebricks (refractory linings)

Common bricks or stone

Blue engineering bricks

Choice of bricks for linings.

abrasion at high temperatures destroys normal bricks or stone. The hot lime and the kiln gases also cause chemical decay of the linings.

There can be no clear advice on the correct material for linings. Some extremely expensive refractory bricks last for many years and their cost may be justified by the lack of interruption of production — particularly if the lime-kiln forms one link in a balanced chemical process.

A suitable compromise must be found between the cost of the units, the inconvenience of shut down for renewal, the life expectancy and the thermal insulation provided. Different materials may be more appropriate for different parts of the kiln.

For the top quarter of a shaft — that is the storage zone and the top part of the preheating zone — the risk of chemical attack is low. Blue engineering bricks are suitable or a hard stone could be used provided it can withstand abrasion and moderate heat. For all the work below this a firebrick will be needed. The usual choice might

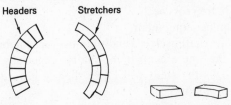

Radial headers and curved stretchers are special bricks for circular structures. Refractory linings are available in these shapes.

be a 40% alumina brick with a good hard and smooth surface. This would be laid on a mortar of fireclay (not lime which would decompose). The joints should be kept as fine as possible as the joints are more vulnerable than the bricks. Some linings are pre-

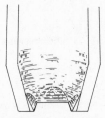

The change from circular section to a square opening at the base of a shaft kiln.

pared as carefully shaped curved and tapered units designed to follow the shape of the kiln and to keep the joints fine.

Sudden changes of temperature can damage the kiln linings and structure. If at all possible, the kiln should be heated very gently up to its operating state.

Kiln linings are usually just 22cm thick but in high efficiency high cost kilns they may reach 69cm thickness at the calcining zone.

6.7 Internal shape of shaft kilns

An ideal shape for a shaft kiln might be a simple cylinder, but for various practical reasons designers have at some time introduced constrictions at almost every level of the kiln. In some cases the constriction gave the required advantage but also introduced an unforeseen disadvantage.

There plainly needs to be some change of shape at the point of discharge and this is frequently brought down to a truncated cone with walls not more than 30° from the vertical. The cone may also change to a square cross-section at the point of discharge to allow simpler mechanical equipment.

The wall of the shaft itself is usually almost vertical, but if there is any danger of the lime sticking in the shaft — especially with clayey limestone feed — the shaft will taper slightly inwards to give a release angle of about 3°.

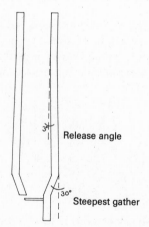

Shape of shaft kilns.

6.8 Tension bands and steel casings

All but the shortest shaft kilns must withstand substantial outward pressures on the kiln walls. These were traditionally resisted by very thick walls, or by adding buttresses, by building behind retaining walls or by building underground.

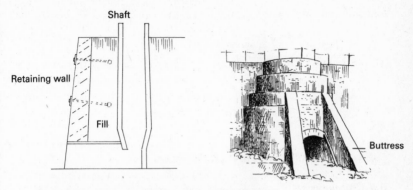

Traditional ways of resisting outward pressures.

A very simple way to overcome the problem is to use steel tension bands around the kiln. The Khadi and Village Industries Commission kilns use bands of 6mm × 76mm at intervals of 800mm up the height of the shaft. Each band is tightened with a bolt and shackle. The height of this shaft was only 4.1m.

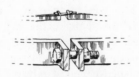

Narrow steel tension band

60mm x 760mm tension bands on KVIC kiln.

The usual way to build tall kilns is with a steel casing. This can provide greater resistance to tension and also helps to prevent leaks through cracked masonry. It is essential to include a compressible insulation material — such as rockwool — between the casing and the brickwork to allow for the different rates of expansion of steel and masonry.

As the inner face of the steel cannot be inspected once the lining is in place it should be well protected against rusting. The usual practice is to set up the casing first and then to build the brickwork inside it. The casings will be expensive. They are often formed of steel sheets about 150cm square and of thickness ranging between 0.6cm and 1cm, perhaps even thicker at the base of large kilns.

Steel casing with covers at inspection holes.

6.9 Chimneys

At any steady temperature the reaction in a lime-kiln will come to a halt unless the carbon dioxide gas can be removed. This means that the partial pressure of the CO_2 must be kept low by introducing a flow of air or steam to dilute the kiln gases. If there is some commercial advantage in gathering the carbon dioxide — as for a sugar beet refinery — the draught may be induced by a fan. However, for small-scale lime-burning it is usual to rely on natural draught. This can be improved by the use of a chimney which can raise the effective height and make the working conditions less unpleasant during loading. A chimney also reduces the irregular burning which wind variations will produce on an open topped kiln.

Simple chimneys may be formed of galvanized iron sheeting. For the 120cm diameter KVIC improved kiln the chimney is formed from an inverted saucer shape carrying a 235cm high pipe which tapers from 26cm down to 20cm. this pipe has a conical cap to keep out rain. The whole chimney is removed whenever the kiln is recharged.

In some kilns the base if the chimney is submerged 1m below the normal surface level of the storage zone at the top of the kiln. The top of the shaft is protected by a cover, but when this is opened for recharging the majority of the kiln gases will still tend to pass up through the chimney.

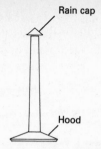

A galvanized iron chimney for a KVIC shaft kiln.

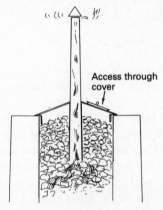

A dipping chimney. Showing the bottom of the chimney well submerged.

The higher a chimney is made the stronger the draught will be. If a kiln is on the quarry bottom a high chimney may be needed to avoid downdraughts, but normally for small-scale production the height will be between 2.5 and 6 metres.

6.10 Roofs
If rain enters a lime-kiln heat is wasted. Both batch and shaft kilns should be well protected. A good roof can also make working conditions easier and safer, but the area at the top of the kiln must be well ventilated.

The experimental batch kiln used by the Transport and Road Research Laboratory in Ghana had a simple corrugated iron roof, but this was too close to the top of the kiln and the sheets decayed quickly in the hot acid fumes.

The roof over a range of kilns in Malaysia is double pitched and staggered in three sections so that the fumes may readily escape between the sections.

A well ventilated roof over a range of kilns.

For a small shaft kiln built in Tanzania the roof was supported off the wooden scaffolding used when the kiln was built.

6.11 Pokeholes and inspection holes

The charge should flow down a shaft kiln without difficulty, but this does not always happen. Sometimes the burnt lime will form an arch across the shaft to block the flow. Again, the material may fuse into vertical columns or 'scaffolding' up the height of the kiln. These problems are especially likely when burning hydraulic limes.

The obstructions can be broken away by poking with an iron rod through one or more of the poke holes in the walls of the kiln. In a kiln in Tanzania poke holes were arranged in a spiral form around the kiln with one hole provided to every five courses of brickwork (about 35cm). Each hole is the size of one brick (12cm × 8cm) so

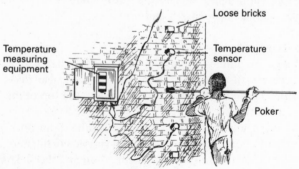

Loose bricks

Temperature measuring equipment

Temperature sensor

Poker

The use of poke holes and inspection holes.

that it can be closed by sliding in a loose brick. The KVIC kiln has poke holes at every 80cm up the height of the shaft.

These holes are also used for other purposes as they provide windows into the kiln. The lime-burner can judge the temperature from the colour of the charge or the holes may be used for electrical temperature sensors (pyrometers) and even for automatic control of the feed.

The poke holes must have airtight doors if the kiln has a steel casing.

Permanent inspection holes may be provided in the calcining zone. These need only be about 50mm diameter and should have permanent covers of mica.

6.12 Access platforms

Platforms will be needed around most parts of the kiln to give access to the pokeholes and inspection points. These should be sturdy and about 1.5m wide to allow the equipment to be handled comfortably. The decking should not be slippery. Handrails will be needed about 1m above the decking and there should be a vertical toeboard about 15cm high. The safety standards for scaffolding on building sites would be an appropriate guide.

A generous and covered platform should be provided at the top of the kiln. If trucks are used for charging a free space of 1.5m should be left between the trucks and the handrail to allow a man to pass a truck safely. Where a charge is stored on the platform this should be supported directly over the kiln wall.

6.13 Loading a flare kiln

All lime-kilns are loaded or stacked to provide an even distribution of heat through the material.

The lime-burner charging a flare kiln may place every stone by hand picking out the biggest stones to arch over the grate and then filling the kiln with stones of diminishing size. Stones of the same size, whether large or small, will pack together leaving up to one third of the total volume free for the gases to pass. When large and small stones are mixed and small ones fill the gaps between the large ones and the passageways for the gases are reduced. If the lime-burner makes the change from large stones to small stones very gradually the passageways remain open. The largest stones need most heat to convert them to lime so they are placed over the fire.

The smallest stones are used where the kiln is coolest at the walls and at the top. In large flare kilns there are openings for loading at several levels in the sides. They are blocked up as the stacking proceeds.

6.14 Loading shaft kilns and ring kilns

The initial charge in a shaft kiln may be placed with care but the subsequent filling is likely to be by tipping in from barrows, dumper trucks or tipper lorries. It is still important to maintain the open network of voids and so it is usual to use stones of one size only. This is normally about the size of a man's fist or perhaps slightly larger. When the stone supply is in more than one grade the smaller stones are sometimes burned in a separate kiln. If they must be burned in the same kiln then the change from one size to another should be made by gradually adjusting the size over several layers of the charge.

In mixed feed shaft kilns the fuel can either be spread in alternate layers or mixed evenly with the limestone. The fuel may tend to block the voids in the same way that undersized pieces of stone would. The method using alternate layers seems to reduce this risk and when firewood is used it is the only practical method.

In some kilns the charge is tipped in more or less automatically. If small particles of coal are mixed with larger limestone pieces then they will tend to separate. The harm from this can be reduced if the charge can be introduced at several points around the top.

One refinement in charging a shaft kiln is to place finer material

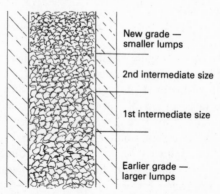

New grade —
smaller lumps

2nd intermediate size

1st intermediate size

Earlier grade —
larger lumps

Gradual adjustment between grades.

Reducing separation by charging through three chutes.

around the circumference of the kiln. This is partly because, as in a flare kiln, the temperature will be rather lower at the edge, but also to help to fill any excessive voids which may form when the stone contracts on calcining. Without this the hot gases can rush up through wide open passageways and reach the top of the kiln before the heat transfer has been achieved. This overburns the lime close to the passage and leaves other stones unburned. This effect is known as 'channelling'.

When ring kilns are stacked the same principles of even gas flow apply. The gases will tend to follow the upper ceiling of the kiln. This is partly overcome by constructing the kiln with drop arches or other means to force the flow downwards but the outer and upper surfaces of the stacks of limestone in the kiln will usually be hottest and any grading of the material should allow for smaller pieces to be at the bottom and at the heart of the stacks with any larger pieces at the outside and especially at the top. Even when underground ring kilns are charged with a grab crane this variation can be partly achieved by loading fine material before the larger grades.

6.15 Firing the kilns

Lime-burning can be treated as an engineering problem in which an apparently simple chemical conversion is to be handled by a well-controlled process. This can be achieved with pure and consistent materials but the problems are really much harder than they first appear. When high fuel efficiency is sought the engineering solutions become very expensive. On the other hand the subject can be treated as an art and conversion is possible with no equipment at all.

Lime-burning as an art can be very successful, but only if the lime-burner is sensitive to everything he controls. Any process can be improved by careful measurement and records of weights and temperatures but there are several indications to help to judge the burning which require no equipment.

The appearance of the smoke or haze at the top of a kiln can be very helpful.

If there is heavy black smoke the fuel has not been burned fully and more combustion air is needed.

If flames burn through the top of the charge the fire is much too high and there is no preheating. Further lime must be drawn and the draught control must be adjusted.

A light haze at the top of the kiln is good. If this were to change to a black smoke with the slightest reduction in air supply then the combustion is correct.

If the air leaving the kiln is quite clear then there is probably too much draught.

Flames   Black smoke

Signs of very inefficient firing.

Another indication is the state of the quicklime when it is drawn from the kiln. Its normal temperature will vary from kiln to kiln but it should be reasonably constant for any one kiln and should be low. Ideally the lime would be only hand hot and at that temperature even slight variations can be sensed without instruments. If it is too hot the fire has probably dropped down the kiln through drawing too much lime out. If it is unusually cold the air intake may be too great. The presence of underburnt and overburnt lumps in the quicklime will also suggest that the calcining temperature is wrong — at least in parts of the kiln. If it can be measured, a rise or fall in the temperature of the exhaust gases will show that the steady operation has been disturbed. In a kiln depending on natural draught the design temperature may be as hot as 200–300°C and this may be difficult to judge although the feel of the kiln structure may give a fair, if sluggish, guide.

The temperature in the calcination zone of the kiln can be judged fairly well if there are inspection holes. At temperatures above 500°C materials radiate visible light and the colour is a characteristic of the temperature:

500°C   Just visible as red light.
700°C   Dark red.
800°C   Just cherry red.
1000°C   Bright red.
1200°C   Bright orange.

Anyone can thus judge which is the hottest part of the charge and with practice it is possible to learn the colour of the 'right' temperature.

The completion of the burning in a flare kiln can be judged by the shrinkage of the charge and by the ease with which a long iron poker can be thrust into the lime.

With only modest apparatus some of these characteristics can be measured and if records are kept there is scope to increase productivity on the scientific basis as well as improving the natural perception of what is right and wrong.

One important point which cannot readily be judged by touch or by eye is the carbon monoxide content of the exhaust gases. The presence of carbon monoxide (instead of carbon dioxide) shows that the fuel is improperly burnt and should have been given more air. This is best detected with analysis equipment but might, inconveniently, be checked by taking a sample tube from above the calcining zone. If the gas burns with a flickering blue flame it is probably carbon monoxide. This gas is poisonous.

CHAPTER 7

Hydration

7.1 Combination of quicklime with water

The English word 'quick' used to mean alive. Quicklime can react so vigorously with water that it was thought to have life in it:

$$CaO \quad + \quad H_2O \quad \rightarrow \quad Ca(OH)_2 \quad + \quad HEAT$$

calcium oxide + water	$\rightarrow$	calcium hydroxide
(quicklime)		(slaked or hydrated lime)

(56 parts) + (18 parts) $\quad \rightarrow \quad$ (74 parts by weight)

If the lime is pure and in a reactive state it may increase in volume by over three times during slaking and release a great deal of heat — up to 1130 kJ/kg.

The affinity of quicklime for water is so strong that great care is needed to keep it dry if it is to be stored. Even though the bulk and weight of an equivalent amount of hydrate will be substantially greater, it is usual for lime to be hydrated at the limeworks and transported in the bulkier form.

Not all limes react as vigorously. The clayey impurities in hydraulic limes slow the reaction. The most impure limes will not slake at all from this lump form. They have to be ground down before use.

Dolomitic limes hydrate in two stages. The calcium oxide component hydrates first, then the magnesium oxide component hydrates much more slowly. Special care must be taken if dolomitic limes are to be used for building. (See Section 3.2.)

The conditions under which the quicklime was formed will also affect the rate of hydration. Limes formed at high kiln temperatures will react more slowly than those formed at lower temperatures. If the quicklime is overburned it will hydrate very slowly. This poses a serious problem when lime is used for plastering by builders. As well as the general expansion of a plaster surface from late hydration, larger particles of overburnt lime may cause intense local expansion which spoils and pits the surface in a very unsightly way.

The answer is to ensure very thorough hydration, perhaps by storing the lime as a moist putty.

Quicklime will combine with water vapour in the air. If the humidity is low this is a slow process, but in high humidity conditions it can be serious. The process is called 'air slaking'. The carbon dioxide gas in the air will also combine with the calcium hydroxide to re-form calcium carbonate. This is one of the ways in which lime mortars harden, but it spoils fresh quicklime. The carbon dioxide does not seem to react with the lime unless there is moisture present.

7.2 Useful forms of hydrated lime

Calcium hydroxide may exist as a dry powder or as a weak solution in water. Between the two extremes there is a range which is suitable for different uses.

Dry hydrate

This is formed by adding just sufficient water to leave the dry powder. In practice a small excess of water would be added, but the heat of the reaction soon dries the material. This is the lightest form of hydrated lime and most suitable for storage and transport. It may be kept in silos, or in good quality paper or plastic sacks. The dry powder could disguise impurities from ground up carbonate core or from overburned particles which have not slaked. A good limeworks will protect its reputation by avoiding these contaminants or working to an agreed specification.

Milk of lime

If quicklime is slaked with a large excess of water and well agitated, it forms a milky suspension known as milk of lime. For many processes, such as beet sugar refining, it is used in this form. The milk may contain impurities, especially from underburnt lime (calcium carbonate) and from overburnt lime (calcium oxide in an unreactive state). There may also be impurities from calcium silicates and aluminates if the original limestone was impure. The active ingredients of the milk will be a concentrated solution of calcium hydroxide in water and a suspension of colloidal micelles. These are agglomerations of calcium hydroxide loosely combined with water. There may also be calcium hydroxide in a crystalline form.

The solids settle out from the milk very slowly and this delay is used in some processes. The larger the agglomerations the lower their densities and so the slower they settle.

Lime water

When the milk settles out, the clear liquid solution left behind is called lime water. This would normally be discarded as it is too weak to be useful. There is a very small specialist use of lime water in the conservation of limestone sculpture.

Lime putty

The solids in the milk of lime gradually settle down and in time the agglomerations absorb more of the water. If the lime water is drawn off the residue is termed lime putty. It is a viscous mass which tenaciously holds the water it has absorbed. Lime putty is one of the finest building materials. It may be stored indefinitely under moist conditions and if left uncovered it forms a crust that protects the heart for many years. The longer it is stored the better it becomes. In time all the overburned oxide slakes in the putty, then there is no possibility of unsoundness in the final work. Lime putty is bulkier and much heavier than the equivalent amount of dry hydrate. For this reason it is not used as widely as it should be.

Dry hydrate may be mixed with water to form a lime putty. If this is stored for a long period it approaches the quality of a putty formed directly from the quicklime, but for the first day or two it has inferior properties of plasticity and water retention.

7.3 Water-burned and drowned lime

One of the more obvious hazards when hydrating lime is that a part of the quicklime may hydrate so slowly that it passes through the process without being converted. In milks and putties this is rarely important, but for dry hydrates it can be a serious defect.

There are two further hazards which are less obvious and not well understood. If water is added too slowly the heat evolved may raise the temperature of the lime as high as 400°C. At any temperature over 100°C (the boiling point of water) there is a risk that the lime will be 'water-burned'. Instead of a powder, an inactive white gritty compound is produced. The chemistry is in doubt, but it seems that a double compound of calcium oxide and calcium hydroxide is formed. This may be as a skin around a core of quicklime. The water-burned lime spoils the quality of a milk of lime for some processes and it is a waste of available lime for all uses.

The third hazard can arise if water is added too quickly. The result is known as drowned lime. Slaking is slow at low temperatures, so that the process starts slowly and then accelerates as heat is generated. If too much water is added to a slow slaking lime it may form a skin of hydroxide. This acts as a barrier between the quicklime and the water and prevents further hydration. As with overburned lime, the hydration may take place later through air slaking. If the lime has been used for plastering, the surface may be spoiled.

There are two distinct techniques of hydration. In one method water (or steam) is added to the quicklime. In the other the quicklime is added to an excess of water. Both methods can be successful.

7.4 Slaking by hand on a platform
This is usually the best method of producing the dry hydrate by hand. The main piece of equipment is a concrete or masonry platform. This would be at a height of about 30cm above ground level to avoid contamination from the surrounding soil and to raise the materials to a comfortable working height. The dimensions are not important, but the top should be level.

Slaking by hand on a platform.

The quicklime is broken into small pieces if the lumps are large and then is spread over the platform in a layer between 10cm and 15cm thick. Water is sprinkled gently over the whole surface in one continuous application. The exact amount will depend on the characteristics of the quicklime, and experience will soon show how much is needed. For a good pure lime the weight of water will be about half the weight of the quicklime. This allows for a fair amount of evaporation. The lime should then be turned over several times with a shovel. The heat which is generated must be conserved and this can be done by covering the lime with hessian sacking, but it must be watched for the first few hours as the sacking could catch fire from the heat of the slaking. The heap should be left for a full day to complete its slaking.

After this the lime should have crumbled to a fine powder, but may contain impurities. Before bagging it should be passed through a sieve. For good quality hydrate a 250 micron (i.e. 60 mesh) sieve is used, but for other uses a coarser sieve may be appropriate. The residue should be discarded or set aside for use in the soil.

Using a coarse sieve for low grade dry hydrate.

If the correct amount of water is added, the lime falls to a dry powder. But if rather too much is used it initially forms a paste. This may still dry out to a powder as it generates more heat and chemically combines with more water.

Using this method the lime is unlikely to be drowned but care must be taken to avoid 'burning'. The two important points are to sprinkle the water gently but continuously and to mix the materials well with the shovel. Any obviously overburned lumps of quicklime should be picked out by hand before slaking.

The method will not work satisfactorily for the more hydraulic (impure) limes as they will not readily break up in a reasonable time.

There is a refinement to separate out the most reactive limes. For this the watering is interrupted when most of the water has been added. The material is passed through a coarse sieve (12–18mm) and the finer material is finished separately from the coarser particles. As the fine materials hydrate more quickly they are likely to be more reactive. Final sieving will still be needed as before.

7.5 Slaking by hand in a tank

Quicklime may be slaked in a tank of water to form milk of lime. This will settle out to give a lime putty. This is the method usually used for soft or very soft forms of lime.

A shallow waterproof tank is needed. It should be about 40cm high and filled with water to a depth of 25cm to 30cm. The bottom of the tank should be covered with quicklime to a depth of about 15cm and the lime should be constantly stirred with a wooden rake. When slaking begins the temperature will rise. As it approaches boiling point further small amounts of quicklime and boiling water may be added and the proportions adjusted to keep the temperature between 90°C and 100°C — that is just below boiling point. The constant stirring must be maintained and all of the lime should be kept below the surface of the water.

For larger batches or for continuous production the tank should have overflows so that the milk of lime can flow out into one or two other tanks at a lower level. These should be 75cm deep below the level of the outlet from the upper tank. The milk settles in these lower tanks. They can be used alternately so that putty may be taken from one whilst the other is filling. A pure lime should be left to mature in the tank for at least three days.

There is a tradition of running quicklime to putty in a pit on a building site. The pit is either dug straight into moist earth perhaps lined with wooden boards, or formed in a bed of the sand which will eventually be used to form the mortar. This is not such a good method of preparing putty as it does not overcome the risks of drowning or burning the lime.

In the tank method there will be no danger of burning the lime provided the lumps are always kept below the water surface. Drowning is a more serious risk, particularly with slow slaking quicklimes. The stirring should prevent drowning and that is why it

A hand slaking tank with two settlement tanks.

is so important to stir constantly, until well after the slaking appears to have stopped. Most of the bulky impurities — especially the residual core — should remain in the top tank which must be cleaned out at the end of the operation. Lumps of noticeably overburned lime should not be put into the slaking tank.

7.6 Mechanical hydration

There are two common forms of mechanical hydrator. In one the quicklime is slaked to milk under violent agitation. This is used only in the 'captive' lime plants serving certain process industries or by very large producers who offer putty for sale as well as the more common dry hydrate.

The more usual type is a hydrator designed to produce dry hydrate by spraying crushed quicklime with a carefully measured amount of boiling water or steam.

Both methods risk the perils of overburning or drowning and once the hydrators are set in operation there is very little supervision. In the milk of lime the impurities are accepted (obviously the machines are designed and set up to perform as well as possible). When good quality dry hydrate is required steps are taken to remove the impurities in an air separator. Shown here is a diagram

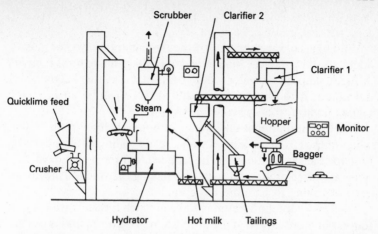

Flow plan of mechanical hydration plant by Fletcher and Stewart Ltd.

of a typical plant to illustrate how much more complicated the simple process becomes when it is automated. The smallest modern hydrators have a capacity of over 40 tonnes per day so they are not available to the small scale limeburner. Capital costs are also high.

7.7 Hydrating impure (hydraulic) limes

Vigorous hydration is a property of the pure limes. The very impure limes will not slake from lumps at all. In between these extremes a typical hydraulic lime will behave as if it were an intimate mixture of pure lime and natural cement. Part of the lime is combined with clayey matter to form the natural cement and the remainder is free. If the proportion of free lime is high enough, the quicklime will crumble when water is added, although more slowly than a pure lime.

It is essential to find out how the lime hydrates by experiment. One common pattern is for a hydraulic lime to hydrate to a mixture of calcium hydroxide (from the free lime), unslaked natural cements in fine particle form and unslaked natural cement in the form of hard nodules called 'grappiers'. Adding further water would slake the fine particles of natural cement which would then set hard. The grappiers are not able to combine with water and must be sieved out. In France these grappiers are then ground finely and either mixed back with the other ingredients to improve the hydraulic qualities (the ability to set hard even under water), or sold separately as grappiers cement.

With limes of this type exactly the right amount of water must be added to slake the free lime and avoid setting the natural cements. A likely amount of water would be 10–15% of the weight of the quicklime. If too little water is added there will be unacceptable contamination from unslaked quicklime.

The hydration might take about 24 hours. If lime can be slaked when still hot from the kiln, or with already boiling water the slaking time for the same lime might be reduced to three hours. Hydraulic limes are often ground finely before slaking to avoid the inconvenience of having to separate the grappiers.

The less hydraulic limes may slake in much the same way as pure limes, but more slowly. If the proportion of clayey matter is low and the kiln temperatures are moderate, there may be no formation of grappiers.

7.8 Hydrating dolomitic limes

When magnesian limestones are burned the dolomitic quicklime does not usually hydrate fully at normal pressures:

$$CaO. MgO + H_2O \rightarrow Ca(OH)_2. MgO + Heat$$
(dolomitic quicklime) + (water) → (normal dolomitic hydrate)

The combined magesium oxide (MgO) can air slake over a long period and cause unsoundness (expansion) in building work. If the quicklime can be hydrated under pressure in an autoclave hydrator it can hydrate fully and be usable for most purposes:

$$CaO. MgO + H_2O + Pressure \rightarrow Ca(OH)_2. Mg(OH)_2 + Heat$$
(dolomitic quicklime) → (highly hydrated dolomitic lime)

A similar hydration is possible at normal atmospheric pressure if the lime is *soft* burned by calcining for a long period at low temperatures (900–1000°C). Hydration on a platform must be with boiling water and the hydration process will be very slow. High temperatures must be maintained throughout. All of this makes the material much harder to handle than a pure lime.

CHAPTER 8
Site Operations

8.1 Site layout and location

Much effort can be wasted by mishandling the materials around a
lime-works. The layout should make the work as light as possible.
At a small lime-works the stone and lime may be handled in
wheelbarrows and this will be easier if there are well made paths
running gently downhill from the quarry to the kiln and from the
kiln to the slaking and despatch areas. Even when mechanical
transport is available a similar arrangement will save fuel and wear
and tear.

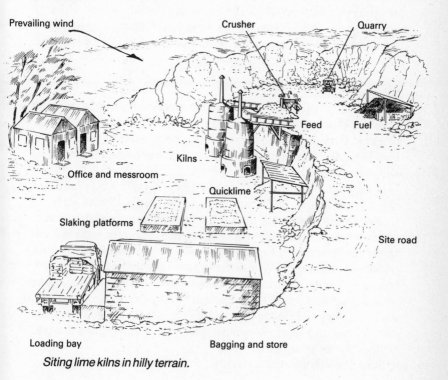

Siting lime kilns in hilly terrain.

If the kiln is at the quarry it should be reasonably close to the workface. When the feed material is delivered to the site by lorry, a well-made access road should be taken right up to the top of the site so that lorries can be unloaded immediately by the kilns.

In the hilly terrain of most limestone districts it is usually possible to find a site where these arrangements can be achieved with only a little cutting and filling of the existing levels. Where the land is flat, as at a coastal lime-works calcining sea shells, it is sometimes possible to design a flare kiln for side loading using one or more ramps.

The access to the site is as important for a small lime-works as it is for a large one. Difficult access will increase the cost of the lime produced. A railway siding or canal would probably not be appropriate for a small works although it may be possible to site the works on an existing waterway, or near a good highway.

Further details of the site layout may be suggested by the prevailing winds. The site offices and messroom will need to be in an upwind position from which the whole site can be seen, yet suitably accessible to visitors entering from the road. Again, it is helpful if the prevailing wind blows across the production line rather than along it and blows any dust away from nearby villages.

8.2 Quarrying

The amounts of limestone or chalk needed to supply a small lime-works are, in quarrying terms, very small. In one small lime-works in England the quarrying operation to feed six lime-kilns is less than full time work for one man who drives both the face shovel and dumper truck.

In the past quarrying was done by hand, but this is very hard and time-consuming work. Weathered and fissured rock may often be picked up using only small levers but heavy hammers and steel bars are needed to work solid rock by hand. Old motor vehicle half-shafts can be forged into jump drills which can be used to drill lines of holes into the rock. Wooden plugs are inserted and then swollen by soaking with water until the rock cracks. In this way blocks can be released which are small enough to manhandle with wedges, levers, rollers and sledges. The blocks would be taken away from the rockface to a safer area on the quarry floor and broken down to smaller pieces with lighter tools.

Similar methods are still used to extract valuable building stone

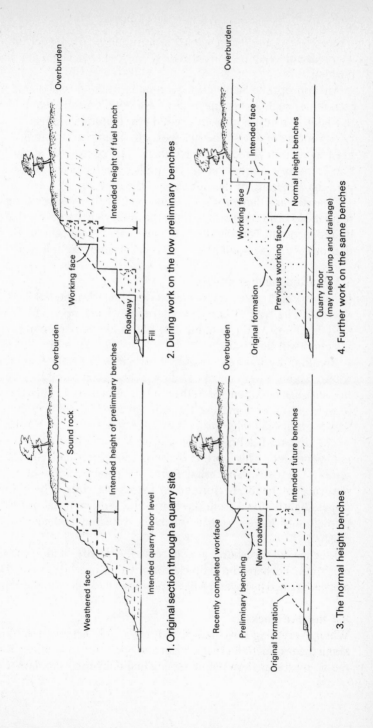

1. Original section through a quarry site

Overburden

Sound rock

Weathered face

Intended height of preliminary benches

Intended quarry floor level

2. During work on the low preliminary benches

Overburden

Working face

Intended height of fuel bench

Roadway

Fill

3. The normal height benches

Overburden

Recently completed workface

Preliminary benching

New roadway

Original formation

Intended future benches

4. Further work on the same benches

Overburden

Intended face

Working face

Normal height benches

Original formation

Previous working face

Quarry floor
(may need jump and drainage)

where blasting could damage the interior of the blocks, but this is a very difficult way to win lumps of limestone or chalk for lime-burning.

An appropriate way must be found to make use of modern techniques such as blasting and to use good crushing equipment. Probably the best approach would be to employ an outside quarrying contractor to visit the site quarry from time to time to drill and blast away a new section of rockface. The contractor might also bring a mobile crusher on his visit. His experience can be used to plan the benching of the quarry and the drainage and access.

Quarries are dangerous and there are likely to be strict safety regulations in all countries. Even if the quarrying work is carried out by independent contractors, the lime-works management should find out the local regulations and see that they are followed.

8.3 Crushing and grading

In earlier chapters the importance of grading the material has been stressed. The type of kiln will determine what shapes and sizes of feed can be used for lime-burning but in all cases preparation or selection will be needed.

Breaking by hand is possible using picks but only economical where labour is very cheap. Unless great care is taken this part of the work can be expensive. If the material is quarried by blasting it is possible to choose and use the explosives in such a way that the debris produces a high proportion of material of a useful size.

The unsorted debris can be picked by hand into the different grades needed. This work can also be done mechanically using a series of screens. The usual form is a rotating cylinder with a sequence of bands perforated to increasing sizes. The cylinder slopes slightly down from the entry end. Any material not falling through the largest screen would be rejected and need to be crushed.

Mechanical crushers are necessarily sturdy and expensive machines. Perhaps the simplest form is that in which heavy hammers are raised by a water wheel and allowed to fall onto the stones.

8.4 Reserve stocks

When everything is working well it is possible for material to pass along the production process from one operation to another. But in the many stages between quarrying and delivering the lime to the user there will often be one or more operations out of action. This may be for routine maintenance, accidental breakdown, staff sickness, or lack of supplies. The reputation of the lime-works will suffer if a break in one item interrupts the production as a whole and delivery is delayed. The remedy is to keep reserve stocks at all stages from raw material, fuel, graded materials, quicklime, slaked lime and bagged products. This all requires extra capital but it is essential if the reputation for reliable delivery is to be maintained.

8.5 Waste materials

The undersized materials from quarrying and crushing cannot be burned in shaft kilns. In a large lime-works they can be burned in a rotary kiln but the small lime-burner has to look for other ways to use them. Part of the spoil may be useful for site restoration after the quarrying. If capital is available to provide the equipment and storage buildings, there is a good potential for crushing or grinding the limestone for use in arable farming as the misnamed agricultural lime. With more elaborate equipment and very considerable care chalk waste may be ground finely as 'chalk whiting' which has many industrial uses, especially as a filler. Some of the larger grades may be useful in road making.

At the hydration area there will be waste from underburned and overburned pieces. The worst pieces will have been hand picked from the lime but the smaller pieces will be left behind when the hydrate is sieved. The old tradition in Europe was to hand pick the good lumps of quicklime from the kiln and sell these for normal building work. The residue, or 'small lime', which contained the overburned lime as well as the ashes or cinders was used to make hard lime concrete floors. In modern practice the waste from the hydration is often sold with the agricultural lime for use on the fields. A modern bad practice is to separate and set aside the denser overburnt particles but to grind the underburnt material and mix it back in with the good hydrate.

CHAPTER 9
Choices to be Made

9.1 Who will buy the lime produced?

The lime-burner must produce the right quality of lime at the right price for his market.

From the outset he must identify the uses to which his lime will be put because this will affect his raw materials and, probably, the location of his works. It will also affect the quality controls needed, the volume of output and the packaging and transport of the product.

For example, any overburnt particles will make a lime unsuitable for plastering buildings but would be acceptable in a sugar beet refinery where the lime is used as a milk. Lime to be used for mortar in buildings could with advantage be an impure hydraulic lime which would be unsuitable for most chemical industries.

It may be possible to offer a range of products for sale, but the operation must be designed at first to match the needs of identified customers.

9.2 Where should the works be sited?

The lime-works may be located at the quarry or near to the place where the lime will be used. An independent works offering lime to a range of customers will usually be sited at the quarry since lime is lighter to transport than the stone or chalk from which it is prepared. But where a single industry, such as steelmaking, uses very large amounts of lime it is usual for the lime-burning to be carried out as part of the overall production process.

In construction projects it would sometimes be appropriate to prepare lime at the site of a major project if the local lime industry could not meet the demand. For example, a major soil stabilization project might require many thousands of tonnes of lime over a period of two or three years. If local suppliers expanded their works to provide this they might be left with too much capacity at the end of the project. On the other hand a lime-burner might offer a contract service moving from project to project.

If the market has been identified then the nature of the necessary raw materials may well determine the location of the lime-works. For large volume production of high calcium lime it is usually necessary to site the works at a quarry where there are thick and even deposits of a pure limestone.

When the general location has been chosen, the possible sites within the area should each be checked for ease of access. The heaviest loads will be the raw material feed for the kilns, but good roads, railways or waterways will also be needed to bring in the fuel and to take away the prepared lime. There must be adequate access for the workmen to reach the site.

Lime-works are dusty and where there is a prevailing wind direction the site should be chosen downwind of any nearby town or village. The nearer the works are to towns or villages the greater will be the effort needed to control this dust and perhaps also pollution from burning the fuel.

9.3 What should be the scale of the operations?

Lime can be burned effectively at almost any scale of operations and the size of individual lime-works may be determined either by the balance of supply and demand or by strategic economic planning.

A large scale of works may be suggested by large individual projects such as a new steelworks or when the raw materials are available only in very limited areas.

Where materials are readily available, a village scale of operations will be desirable as this places the least stress on overall transport needs and gives the greatest sense of involvement to those employed. Indeed, it will be possible to organize a small lime-works as a partnership.

A deliberate choice of scale must be made and this will be based on an estimate of the likely local demand for a good product.

The lime-works should also be designed to accommodate an expansion of perhaps two to five times the initial output as new markets are developed. This is not just for export to other areas but to cover new uses in the immediate area.

The presence of the lime-works may stimulate the use of lime technology for water and sewage treatment or other lime consuming industries. Or again, in an area where lime may initially be used in building only for masonry mortar, the use may be expanded to cover external renderings, limewashes, mass concrete and block-making. To make allowance for this sort of expansion it will be important to choose an inherently flexible system of production.

9.4 Should the production be seasonal?

Lime-burning is possible all year round, although it becomes diffi-cult during periods of very heavy rain. As hydrated limes will store safely, it would be possible to operate a lime-works at times of the year when the labourers are not needed in agriculture or perhaps when building workers cannot be fully employed in winter. The implication is that the capital equipment of the lime-works would be lying idle at other times. The cost of this (the interest charges on capital employed and the maintenance of the plant) must be bal-anced against the benefit of avoiding unproductive time in agricul-ture or in the building trades. It seems likely that only the smallest plants with very low capital costs could make this balance worth-while. On the other hand lime kilns need to be run down for periodic maintenance and it is quite possible to plan for this to be done over perhaps a four to six week period at harvest time.

9.5 What degree of refinement is appropriate?

Those responsible for the lime-works should be able to understand every part of the operation they are controlling. Ideally most of the people engaged on the work would also have a good general understanding of what is being done.

An early decision to be made is whether to choose a method relying heavily on hand labour or to invest substantial sums in mechanical aids. Elaborate mechanical equipment increases the dependence on outside expertize for installation and maintenance. If extensive hand labour is chosen the employees must still be given the best available equipment. Although large scale usually suggests a high degree of automation it should be remembered that ring kilns can give high output and high thermal efficiency from a labour intensive lime-works.

When the highest degree of refinement will normally be sought is in achieving the most effective use of the fuels used. In the past this depended almost entirely on the skill and experience of the lime-burner but in a modern automated process much of the judgement may be made with the aid of a computer. It seems curious to raise this in a work aimed specifically at small scale lime-burners but there may soon be a place for a small computer in even the smallest lime-works.

In certain countries there are national standards defining the quality of lime which should be marketed. In the absence of such formal standards the lime-burner must find out what quality is appropriate for this customers' particular uses and methods of testing must be chosen to maintain this quality. Advanced testing methods are only appropriate for very demanding customer requirements. When the standards have been established, staff training must be arranged to make sure that they can be consistently maintained.

Long-term success will depend on the ability to adapt to growth and contraction and to changes in the work itself. A lime-works based on small essentially simple equipment can more easily be adjusted than one which depends on large, expensive or overcomplicated equipment. The lime-burner must remain master of his tools and not become a mere operator.

9.6 Should a batch process of continuous production be chosen?
For the very smallest annual outputs there is no choice. Some form of batch production will be essential.

Continuous production methods have an inherent advantage of better fuel efficiency and this will always be desirable for the larger lime-works.

At the village scale continuous production is possible with a small shaft kiln but the present equipment is rather inflexible at this scale. It is sometimes found that initial demand and available working capital simply cannot justify continuous production and so the shaft kilns are used ineffectively.

Some of the kilns described in Chapter 6 operate in a batch manner but obtain advantages of heat recovery through a short run of continuous production. The basis is the method of the climbing kilns and the modern development from this is the inclined chimney kiln.

Because of the fuel advantage it is natural to choose the continuous production kiln wherever possible but at the small scale this is often a mistake. If it seems that continuous production may not be maintained it would be better to set up a good batch production method rather than to use the continuous production equipment ineffectively.

9.7 What methods of kiln construction should be used?

Large-scale lime-kilns may be bought from experienced manufacturers and there is no technical reason why small-scale kilns should not be available in much the same way.

But there are good reasons for adopting methods which rely on locally available materials and skills. Very good kilns can be built using only conventional masonry construction, but if good metal-workers are available it will sometimes be better to build a steel cased kiln. By helping other local industries this encourages goodwill towards the lime-works. The masons who construct the kiln may become valued customers for the lime produced.

If a kiln is constructed by local men it will usually be possible for the same men to undertake the repairs later on. This can save valuable time if it avoids delays waiting for spare parts from abroad.

If those building the kiln are not familiar with tall structures it may be necessary to bring in an engineer for guidance. This will also be necessary if the kiln is not founded on level bedrock or firm gravel.

Although high quality kiln linings are not essential, the best refractory bricks will last many times longer than common bricks. If good refractories are not available locally it may well be worth the cost and inconvenience of importing them from far afield to get good overall performance from a shaft kiln.

CHAPTER 10
Evaluation

10.1 The aims must be defined
From the outset it must be known what is expected of the lime-works.

The aim may be very simple — to provide a good profit on the capital invested without transgressing any local legislation. If this is the case the evaluation will be very simple.

At the opposite extreme it may be the intention to produce a certain quantity and quality of lime without particular concern for the cost. Again, evaluation would be simple.

In most cases there will be a subtle blend of environmental, social and economic factors. Whatever the blend is it must be clearly understood from the beginning. If the enterprize seems to be failing then either the project must be adjusted in some way or the objectives should be redefined.

10.2 The effective use of materials and fuels
The cost of all fuels is now so high that any misuse of fuel will quickly be apparent as an economic penalty. The effect of the misuse of the raw materials may emerge more slowly but in the long term it will be the premature exhaustion of supplies. There will also be some immediate penalty since the quarrying costs will be raised if the material is wasted. To guard against this, detailed records must be kept and from year to year improvements must be made. The lime-burner must always remain aware of the need to reduce the fuel used to prepare his lime.

Maintaining the necessary quality is closely related to this. Both overburnt and underburnt lime represent a waste of material and fuel. Good training and alertness will save both.

In this part of his work the lime-burner could be helped by visiting teams from an academic unit — perhaps a university or a national trade association. The academic 'flying squad' could bring more expensive test equipment for detailed analysis of the kiln and could introduce the lime-burner to recent advances in research and development. The detailed records of the operations would help the visitors to judge any improvements following their advice.

10.3 Effective use of labour

Whenever the choice haas been made to undertake a particular task by manpower rather than machine power the labour must be used as effectively as possible. To waste manpower is to insult human dignity, so labour should never be cheap, and it very rarely is.

There are good and bad ways to do every job whether it is maintaining complicated equipment or pushing a wheelbarrow. Every task must be examined and done as well as possible.

There is great value in training employees to do as many of the different tasks as they can. This makes their working lives more interesting and makes it possible to divert resources between one activity and another to overcome any bottlenecks in the production. It also reduces the difficulties which can arise in small enterprises when key members of staff are sick or on leave.

To work effectively the men must have the best tools. Whether this means wheelbarrows or lorries there will be a need for capital to complement the labour. Is the balance of capital and labour right?

10.4 Profitable return on investment

Whatever the blend of aims, a satisfactory return on investment is likely to be a major requirement. Whilst the lime-burner may be able to control and improve the economic performance of the lime-works itself he will have less control over some of the other factors which will affect the return on invested money. For example, changes in interest rates will affect both the cost of any money he has borrowed and the expectation of investors who may see higher possible returns in other areas.

It is not enough to make a consistently good and well priced product if, through lack of advertising or salesmanship, the potential customers are unaware of it. Nor is it sufficient to sell all that has been made if the invoicing is slow and the customers are bad payers. The lime-burner must be skilled not only in the art of his industry but in his commercial activities. This will almost certainly mean further training.

Even a sound industry with a good commercial presentation can still be frustrated by sudden changes in taxation or subsidy or by the loss of a means of transport. If the industry is to be judged a commercial success all of these impediments must be overcome.

10.5 Stimulation of other activities

In a completely free market economy the industry will have been judged on the basis of its profitability. To some extent this will have been influenced by the possible stimulation of complementary industries, but only in so far as they provided profitable outlets for the sales of lime. In a fully planned or a mixed economy the stimulation of other industries will be a major concern. It may be possible to set deliberate targets, but it is more likely that there will be general expectations and the actual progress will be observed from time to time.

It is of course true that any stimulation of the construction industry will come more from the needs for new buildings than from the availability of new materials. But there are other considerations which might play some part in a decision to go into lime-burning. For example, it might be asked whether the availability of good lime can raise the quality of buildings, and whether the lime can free the local builders from the need to import cement from elsewhere? Can the availability of lime encourage new activities in tanneries, paper-making or sugar refining in the area? Can public health benefit from improved waste and sewage treatment? Such a range of possible benefits is itself a clear indication of the importance of this remarkable material.

CHAPTER 11
Glossary

Agricultural lime Any lime used for soil conditioning, but usually used to describe ground chalk or limestone, which is not lime at all.

Air-slaked lime The degenerate product formed naturally when quicklime is stored in moist air. It is a powdery mixture of oxides, hydroxides and carbonates.

Amorphous Not crystalline. Having no definite form or shape.

Amortization The gradual repayment of borrowed capital.

Anaerobic In the absence of free oxygen.

Aragonite The mineral form of calcium carbonate ($CaCO_3$) with an orthorhombal crystal structure.

Arenaceous limestone Limestone containing sand.

Autoclave To heat materials under pressure in a thick walled vessel to temperatures above boiling point, or the vessel itself.

Available lime The oxides or hydroxides of calcium or magnesium which are available to enter into a desired chemical reaction in a particular process.

Batch production Any production method in which the charge is loaded into the kiln, converted to quicklime and is then all unloaded at one time.

Beneficiation of ores The separation of ores into valuable components called concentrates and waste material called gangues.

Blue lias lime An emminently hydraulic lime formerly made in many places in England all lying along a narrow belt stretching from Lyme Regis to Humberside. See also **White lias Lime**.

Burning See **Waterburned lime**.

Calcareous Containing chalk or other forms of calcium carbonate or containing limes.

Calcination In this context the conversion of carbonate to lime, but the word has a much wider meaning including the conversion of metals into their oxides by strong heating.

Calcite (adjective **Calcitic**) The mineral form of calcium carbonate which has a rhombohedral crystalline structure. It occurs naturally as Iceland Spar.

Calcium (Ca) A soft white metallic element which tarnishes rapidly in air.

Calcium carbonate ($CaCO_3$) A solid which occurs naturally as chalk, marble, calcite and the many different forms of limestone. It is also the main constituent in corals and sea shells and a constituent of bones.

Calcium silicate bricks or **sand-lime bricks** Building bricks made of sand with 7–10% lime. The units are moulded at high pressure and cured in an autoclave. Several stable calcium silicate compounds are formed and the bricks are hard and durable.

Calorific value The quantity of heat produced by unit mass of a fuel on complete combustion. The SI unit is joules per kilogram (J/kg).

Carbide lime The calcium hydroxide produced as a by-product when acetylene is generated from calcium carbide.

Carbonate Any salt of carbonic acid. Used in this book to describe calcium carbonate ($CaCO_3$) and dolomite ($CaCO_3.MgCO_3$) used as the feed for a lime-kiln. This includes chalks, limestones and any other form of calcium carbonate such as coral or sea shells.

Carbon dioxide (CO_2) Carbonic acid gas. A colourless gas with a faint tingling smell and taste. It occurs naturally in the atmosphere. It is generated in lime-kilns partly by the combustion of carbon based fuels but mainly by the decomposition of calcium and magnesium carbonates.

Carboniferous Literally means 'bearing carbon' but the geological description 'carboniferous' means that the sedimentary formation as a whole bears carbon, perhaps coal. The individual samples of carboniferous limestone usually contain no carbon.

Cement Generally means a bonding material and in building means a quick setting binder used between sand particles in a mortar or concrete.

Cementation index (CI) One of the methods used to categorize hydraulic limes. It gives an indication of their likely early setting properties based on the proportions of certain constituent compounds or groups of compounds.

$$\text{Cementation index} = \frac{2.8 \times S + 1.1 \times A + 0.7 \times F}{C + 1.4 \times M}$$

where $S = \%SiO_2$, $A = \%Al_2O_3$, $F = \%Fe_2O_3$, $C = \%CaO$, $M = \%MgO$.

Cement rock or cement stone A naturally occurring rock or stone containing the right balance of calcium carbonate and suitable clayey matter to calcine into a strong natural cement or even into a Portland cement.

Chalk A common form of calcium carbonate with a very fine structure.

Charge Used here to describe the carbonate (and also fuel if in a mixed feed kiln) which is fed into a lime-kiln.

Coarse stuff A mixture of coarse sand and lime used in building for the first coat of plastering or for mortar for masonry. A good building practice is to store the coarse stuff whilst the lime 'fattens'.

Composite mortar or **compo** A building mortar containing both lime and cement as the binder for the sand.

Concrete A structural material formed by mixing carefully proportioned sand, stones, water and a binder. The binder is usually Portland cement but may be hydraulic lime or lime and a pozzolan.

Coral A natural form of calcium carbonate formed in warm seas by secretion from marine polyps.

Core The carbonate at the heart of an underburned piece of quicklime.

Core samples Samples of rock from within a formation obtained with a special hollow drill.

Deadburned dolomite A chemically inactive form of dolomitic quicklime used for refractory linings.

Deadburned lime Calcium oxide formed at excessively high kiln temperatures. It has a dense physical structure which does not allow it to hydrate readily under normal conditions.

Decrepitation Collapse of a lump by sudden and general cracking during calcination.

Dissociation The reversible decomposition of the molecules of a compound. In this context, the separation of carbon dioxide from the carbonates to form lime.

Dolomite ($CaCO_3$. $MgCO_3$) The double carbonate of magnesium and calcium.

Dolomitic limestone, magnesian limestones Limestones containing both dolomite and calcium carbonate.

Draw kiln One of the names for a continuously operated shaft kiln. Lime can be drawn from the kiln at the bottom and fresh charge added at the top whilst the kiln is burning.

Drowned lime Lime which has been spoiled in slaking because it has failed to reach the necessary temperature for a satisfactory reaction between the water and a naturally slow-slaking (unreactive) quick-lime. A skin of lime putty seals the lump preventing further fresh water from reaching the quicklime.

Dry hydrate of lime ($Ca(OH)_2$) Calcium hydroxide, or slaked lime, in its dry powder form.

Eminently hydraulic lime Lime from an impure clayey carbonate which would set under water within two to four days. The active clay content of the parent limestone would be about 18–25% and the cementation index about 0.7–1.1.

Exhaust gases or exit gases In this context the gases leaving a lime-kiln. These include the products of combustion and of dissociation, also steam and any air passing through the kiln.

Exothermic reaction A chemical reaction which generates heat. For example, the slaking of quicklime.

Fallen lime Air-slaked lime.

Fat lime A relatively pure lime and not a hydraulic lime. Especially one that would yield a good workable lime putty for building uses.

Fattening The slow absorbtion of water into a lime putty. This literally plumps it up and makes it more plastic for building uses.

Feebly hydraulic lime A hydraulic lime which would set underwater in more than 20 days. The active clay content of the parent limestone would be less than about 12% and the cementation index would be about 0.3–0.5.

Ferruginous limestone A limestone containing iron compounds.

Fine stuff A mixture of finely sieved sand and lime putty for plastering.

Flaggy That which can be split, or is split, into thin sheets.

Flocculation The coagulation of finely divided particles into fewer and larger particles.

Flux A substance added to others to assist fusion.

Fossils The remains of organisms contained with the Earth's crust.

Fossiliferous limestone Limestone containing fossils.

Free lime That part of the calcium or magnesium oxide in a hydraulic lime or cement which is not compounded with aluminates, silicates or iron oxides. If there is too little free lime, the lump will not break up naturally on hydration and the quicklime would have to be ground.

Grappiers Lumps of clinker which are formed when certain hydraulic limestones are calcined. These are screened out from the remainder of the lime and ground to a fine powder which is either mixed back with the lime or sold separately as grappiers cement.

Green firewood Green is used here to mean fresh and not seasoned. It is not necessarily describing a colour.

Grey lime, Greystone lime Semi-hydraulic limes. Their colour is often a warm grey.

Ground lime Quicklime which has been ground down to a certain size. This would be specified and might be all to pass a No. 8 sieve and 40–60% to pass a No. 10 sieve.

Gypsum ($CaSO_4.2H_2O$) Hydrated calcium sulphate which loses 75% of its water of crystallization to form plaster of Paris when heated to 120°C.

Hand picked lump lime Quicklime which has been selected lump by lump from the run of kiln quicklime and should thus not contain seriously overburned or underburned lime. The description Best Hand Picked or BHP was frequently specified.

Hardburned lime Quicklime which, though not visibly overburnt, has been calcined at a high temperature and is slow to hydrate.

High calcium lime, high calcium limestone High calcium lime is that prepared from high calcium limestone. High calcium limestone must contain at least 95% calcium carbonate. This specification thus distinguishes magnesia (MgO) as an impurity.

High magnesium dolomite A dolomitic limestone containing more than 43% magnesium carbonate. In this specification the free calcium carbonate is thus distinguished as an impurity.

High purity carbonate rock Rock in which the calcium and magnesium carbonate content together exceed 95% of the total weight. In this specification the sandy and clayey materials are distinguished as impurities.

Humic Adjective derived from humus, decayed vegetable substances.

Hydration Slaking. The combination of water and quicklime to form hydroxides. This is an exothermic chemical reaction.

Hydrated lime The dry powder obtained by slaking quicklime with enough water to form the hydroxide. The hydrate may be of a high calcium lime, a magnesium lime, dolomitic lime or of some form of hydraulic lime.

Hydrated hydraulic lime A lime with active clayey impurities which have combined with part of the quicklime to form natural cements and which the remaining free lime has been hydrated from the oxide to the hydroxide form. This is used in building and is in powder form.

Hydraulic limes Limes which will set, even under water, and with which hydraulic engineering structures (harbours, canal locks, bridges etc.) can be built. These limes are formed from impure limestones and the setting properties come from the compounds of silica, alumina and ferric oxide with lime and magnesia.

Hydroxide A compound in which one of the hydrogen (H) atoms in water (H_2O) has been replaced by some other atom or group of atoms; a compound containing the hydroxyl group (—OH).

Igneous (rocks) Formed by the action of fire.

Kiln Any structure or chamber in which materials are heated. Temperatures would usually be greater than those in any oven and lower than those in any furnace. Different forms of kiln are used for lime-making, pottery, brick-making and other processes.

Kiln dust Dust drawn from the bottom of a kiln. From a mixed feed kiln this would include coal ash, air-slaked lime, fine particles of quicklime, fine particles of overburned lime and other calcined impurities. The material was used traditionally for making concrete floors.

Large lump lime Quicklime with lumps sized over 200mm. Hand picking to select lumps is simpler with large sizes.

Lias lime, lyas lime See **Blue lias lime**.

Lightburned lime Lime which has been calcined at a relatively low temperature and is thus more reactive than other limes.

Lime In this context the term lime includes all oxides and hydroxides of calcium and magnesium. The quicklime and slaked lime forms are thus included but the carbonate forms are not included. In other context lime is: a viscous material (such as bird lime); ground limestone misnamed agricultural lime; and as a verb, the act of spreading a viscous material or applying agricultural lime.

Lime ashes As kiln dust, but especially when wood ash or coal ash from the hearth of a flare kiln are included. The ash has a pozzolanic action.

Lime-burning Converting calcium (and perhaps magnesium) carbonate to quicklime in a kiln.

Lime concrete A building material cast from aggregate (usually sand and stone) in a matrix of hydraulic lime or lime and pozzolana, but not using Portland cement.

Lime mortar Sand bound in a matrix of lime or lime and pozzolana (but not Portland cement) and used for laying bricks and blocks in buildings.

Lime plaster A mixture of water, lime, sand and sometimes hair which is applied to walls and ceilings in a plastic state before drying to give a hard smooth surface for decoration. In good quality work the plaster is applied in three coats (layers) using progressively finer materials. The first coat uses 'coarse stuff', the second coat uses 'fine stuff' and the finishing coat uses lime putty. In two coat work the finishing coat is the 'fine stuff'.

Lime pit, maturing pit The pit in the ground on a building site where lime is slaked and stored as putty, often in a bed of sand.

Lime powder Dry hydrated lime.

Lime putty Slaked lime stored in an excess of water to fatten up. This process also enables less reactive particles to be hydrated. In Roman times the putty for plastering had, by law, to be stored for three years.

Lime water A clear solution of hydrated lime in water. It is used for the conservation of limestone sculpture and as a chemical reagent.

Limestone Any rock or stone whose main constituents are calcium carbonate or a combination of calcium and magnesium carbonates.

Limestone lime Lime prepared from limestone rather than chalk or other materials. For many hundreds of years this had the reputation of being 'better' than chalk lime.

Limewash Paint prepared readily from lime with or without various additives. It is suitable for use on walls.

Lining The refractory layer on the inner face of the shell of a limekiln. It must survive abrasive contact with the burning lime. Also, to insert such a lining.

Lump lime Quicklime in substantial pieces, usually in the size range 65mm to 300mm.

Magmatic Adjective derived from magma, the molten mineral matter deep in the Earth.

Magnesia (MgO) The oxide of magnesium.

Magnesian limestone Dolomitic limestone containing between 5% and 40% magnesium carbonate ($MgCO_3$).

Magnesium (Mg) A metallic chemical element.

Marble In this context marble is a stone with a very high calcium carbonate content which has undergone metamorphosis under the action of pressure, and possibly heat, giving it a new wholly crystalline structure. It takes a decorative polish. In other contexts, any decorative stone which can be polished.

Marl Unconsolidated deposit of fine carbonates and clay or earth. Used for improving soil in agriculture and occasionally for lime production.

Masonry Building elements with bricks, blocks or stones. In some contexts the word is used to exclude brickwork.

Metamorphic rocks Rocks whose structure has become less or more crystalline under the action of heat and pressure.

Maturing In this context means the fattening of lime putty.

Micron A unit of length. One micron was one millionth of a metre. Now renamed a micrometre.

Milk of lime A white, milk-like, suspension of hydrated lime in water.

Mixed feed The charge for a shaft kiln when the carbonate and fuel both enter at the storage/preheating zones. Also used to describe this method of operation. The alternative is some form of separate firing.

Moderately hydraulic lime A hydraulic lime which would normally set under water in between 5 and 20 days. The active clay content in the parent limestone would be about 12–18% and the cementation index would be about 0.5–0.7.

Mortar A mixture of water, sand and a binder used to joint bricks, blocks and stones in building.

Mountain limestone, mountain lime The massive, thick bedded deposits of limestone in the carboniferous limestone series. These are often very pure and of great commercial importance. Mountain lime is a fat lime.

Natural cement A quick setting hydraulic cement prepared by calcining naturally occurring cement stones such as septaria. These stones contain an appropriate clay content and no blending is needed before calcining at only a little above lime-kiln temperature, not to the temperatures used to form clinker for portland cement. They contain insufficient free lime to break up on hydration and must be ground up for use. One common name for natural cements was Roman cement.

Ordinary Portland cement (OPC) An artificial hydraulic cement prepared from clay and chalk, marl or limestone at kiln temperatures high enough to form a clinker which must be ground finely. It would conform to appropriate clearly specified standards. This is the normal cement used for construction.

Overburnt lime Quicklime which has been calcined at too high a temperature and which will not slake readily. In extreme cases it shrivels up and has a wizened appearance.

Oxide A binary compound with oxygen.

Parker's cement A natural cement patented in England in 1796 by James Parker.

Partial pressure The pressure exerted by a particular gas in a mixture of gases.

Pebble lime Quicklime of a certain size. It is between about 6mm and 65mm.

Periclase A dense and stable form of magnesia (MgO).

Petrology The specialist geological study of rocks.

PFA Pulverized Fly Ash. A pozzolan prepared from the ash of certain coal fired electricity generating stations.

Pitting or popping Local failure of plaster some months or even years after application. A particle of overburnt or poorly slaked lime expands on late hydration and may cause a small explosion as the cone of plaster in front of it breaks away. This can be avoided by air separation of the hydrate or by fine sieving (No. 50 mesh) or by extended storage of the lime as lime putty.

Plasticizer In this context, an additive to make a mortar or concrete more workable (or plastic).

Poke holes Small holes left through the shell, lining and casing of a shaft kiln to allow the contents to be poked. The holes have other uses.

Portland cement Artificial cement so named because it was as strong as Portland stone (then a highly prized building stone). There are various standard specifications available.

Pozzolana An Italian volcanic ash from Pozzuoli near Naples. It has been used since Roman times to produce hard and waterproof lime concretes and mortars.

Pozzolan, pozzolanic Pozzolans react with limes to assist their setting properties. Natural pozzolans of volcanic origin include pozzolana from Italy, trass from the Rhineland and santorin from the Greek Islands. There are other exploited deposits in France, the Azores, the Canary Islands and Tanzania. Deposits may be expected in any region which has had volcanic activity. Artificial pozzolans include PFA, blast furnace slag and several fired clays including the Indian material, surkhi. All pozzolans contain clays which have been rendered active by heat, but there is a very wide variation in their properties.

Preheating zone The part of a kiln where the charge is heated up to just below its dissociation temperature.

Producer gas A fuel prepared locally in a gas producer. It has a low calorific value and contains a high proportion of carbon monoxide.

Pulverized fly ash A fine pozzolan obtained as a waste product from certain coal fired electricity generating stations. Sold as PFA and used for grouting and for partial replacement of cement in certain concretes.

Pulverized lime Quicklime of a size which will all pass through a No. 20 sieve and of which around 90% will pass through a No. 100 sieve.

Pure lime Lime with negligible impurities, particularly without clayey impurities which would make it hydraulic. Fat Lime.

Quicklime Lime which has not been slaked. A calcined material whose major component is calcium oxide (CaO) or calcium oxide in natural association with magnesium oxide (CaO.MgO) and which is capable of being slaked with water. It is often described by its size grading, for example pebble lime and lump lime.

Rank In the context of coals rank is the indication of quality and of the degree of metamorphosis. Coal is formed by the conversion of decayed vegetable matter to carbon and the further this change has advanced, the higher the rank will be.

Reactivity of lime The ability of lime to combine quickly in chemical reactions. This can be seen immediately in the slaking reaction. It is partly dependent on the parent limestone and largely dependent on the temperature and duration of calcination. Reactive limes have porous structures with high surface areas. Softburned limes are reactive. Hardburned limes are less reactive. Deadburned and overburned limes are very unreactive.

Refractory Able to resist decay under extreme conditions in a kiln or furnace. Also, a brick which has refractory properties and would be suitable for lining a kiln.

Roman cement Natural cement. A description used widely in the nineteenth century.

Run of kiln quicklime Quicklime in the state in which it is drawn from the kiln including any contamination with ash, underburnt or overburnt pieces.

Running kiln A shaft kiln, especially an unsophisticated one. The material runs (albeit very slowly) through the kiln.

Running lime to putty Slaking lime, especially on a building site either directly in a lime pit or to milk of lime which is then sieved into the maturing pit.

Sand-lime bricks Now generally called calcium silicate bricks.

Scaffolding In this context the formation of columns of fused material within a shaft kiln. These also develop into arches and prevent the charge from falling correctly downwards through the kiln. They make poke holes worth while.

Sedimentary rocks Rocks which have been formed by materials deposited in water and which are generally laid down in distinct layers.

Semi-hydraulic lime A lime with slight hydraulic properties which might fall into the categories of feebly hydraulic or moderately hydraulic lime.

Septaria Cement stones occurring in the mud of the Thames estuary in England and in other places. They are sometimes in the form of squat seven sided prisms.

Shell In this context either sea shells which may be used as a source of calcium carbonate, or the wall forming the structural enclosure of a lime-kiln.

Shell lime Lime prepared from sea shells. Generally a pure lime.

Slaked lime Lime which has been slaked with water to the hydroxide form. It may be in the form of dry hydrate, putty, slurry, milk of lime or even limewater. It may be pure or hydraulic and calcitic or magnesian. $Ca(OH)_2$, $Ca(OH)_2.MgO$, $Ca(OH)_2.Mg(OH)_2$.

Slaking The addition of water to quicklime.

Small lime When it was the normal practice for builders to choose 'hand picked lump lime', the remainder of the quicklime from a batch, including dust and ashes, was sold as 'small lime' at a lower price. It was used for flooring and for foundations.

Softburned lime Lime burned at a low temperature and thus in a reactive state with a low density and high surface area. A valuable product.

Solidburned lime The opposite of soft-burned lime. It has been produced at a relatively high temperature for ease of production, it is less reactive and less valuable but not necessarily less profitable.

Soundness and unsoundness Soundness is one of the properties required of a lime plaster. If a plaster is unsound it contains poorly hydrated material in a finely divided state. In time the fine particles hydrate and expand causing a general expansion of the plaster leading to serious cracking, especially between the plaster as a whole and the wall to which it was applied. Soundness is measured by an expansion test.

Stratigraphy Study of the relationships between layers or strata.

Stoichiometric mixture A mixture of reagents (such as fuel and oxygen for its combustion which are in exactly the right proportions for their combination into a compound (the product of combustion) leaving no surplus of any of the reagents.

Stone lime Lime from limestone rather than from chalk or any other source.

Trass or tras A natural pozzolan from Germany. It was exported to England down the river Rhine through Holland and was thus called Dutch Trass.

Ultra high calcium limestone Limestone containing more than 97% calcium carbonate ($CaCO_3$).

Underburned lime Carbonate which has passed through the kiln without dissociating completely thus leaving a core of carbonate within the quicklime.

Unslaked lime Quicklime.

Vertical kiln A shaft kiln. Generally used to distinguish a shaft kiln from a rotary kiln in the United States.

Volume yield The volume of putty of standard consistency obtained per unit weight of quicklime. Measured in litres per kilogram. This is a useful method of quality control.

Waterburned lime Lime which has failed to slake correctly through the formation of a protective skin at high temperatures. This is probably a double compound of calcium oxide and calcium hydroxide ($CaO.Ca(OH)_2$).

Water limes or water building limes Hydraulic limes. Limes which can be used for masonry or concrete to be submerged under water.

Water retentivity The ability of mortar to retain water against the suction of bricks or stone and against evaporation in hot climates. This is needed to achieve a good bond between the mortar and the masonry units. It is closely related to the plasticity or workability of a mortar and can be measured by the flow of a mortar sample when tested on a standard flow table before and after application of a specified suction.

White chalk lime A white and non-hydraulic lime as opposed to grey chalk lime.

White lias limestone A relatively pure lime from the lias formation and not the emminently hydraulic blue lias lime from the adjacent blue lias limestone.

White lime A non-hydraulic lime. Compare this with grey lime.

Whitewash Either a thin lime wash or a paint made from whiting, size and water.

Whiting Finely ground chalk with a wide range of industrial uses.

Workability In this context, plasticity. The ability of a mortar to spread smoothly and lightly.

CHAPTER 12

Where to Find Help

12.1 Reference materials
General textbooks
There were two English textbooks which dealt at length with the subject of lime-burning, although neither dealt specifically with the problems of small-scale lime-burning:

1. Searle, A.B. *Limestone and its products, their Nature, Production and Uses.* London, 1935
2. Knibbs, N.V.S. *Lime and Magnesia* London 1924

Both dealt with the subject in a practical way, but they have been out of print for many years.

There is an American textbook still in print. This deals with many of the theoretical problems, but the practical sections relate entirely to large scale commercial production in the United States and a few advanced industrial countries:

3. Boynton, R.S. *Chemistry and Technology of Lime and Limestone* New York. 2nd Edition 1980

These three books were written for industrialists, works managers and specialist students. A much lighter work was published in 1963 and this would be more suitable for use in school sixth forms and technical colleges:

4. Stowell, F.P. *Limestone as a Raw Material in Industry.* London 1963

The most recent general work on small scale production is the study from GRET which outlines production methods and uses for the construction industry:

5. Groupe de Recherce et d'Echanges Technologiques *La Chaux: sa Production et son Utilisation dans l'Habitat*

A brief but valuable introduction to the subject is available in the

'Overseas Building Notes' series from the Building Research Establishment in England.

6. Bessey, G. *Production and Use of Lime in Developing Countries* Overseas Building Note 161. British Research Establishment Garston, 1975

References for Chapter 1
The nature of lime is covered by Boynton (3), but perhaps the most explicit description of the range of materials and their uses in building works is given in one of the early publications from the Building Research Station in England:

7. Cowper, A.D. *Lime and Lime Mortars* Department of Scientific and Industrial Research, Building Research Special Report No: 9,. H.M.S.O. London 1927

Boynton (3) and Searle (1) both outline the industrial processes in which lime is used. A little more detail is given in a booklet from the N.L.A. in the United States:

8. National Lime Association *Chemical Lime Facts* Baltimore 1951. Third Edition, 1973

Explanations of individual industrial processes will only be found in the textbooks dealing with those industries such as:

9. Hugot, E. *Handbook of Cane Sugar Engineering* Amsterdam, Oxford, New York 1972

The role of lime as an alternative to ordinary Portland cement can be seen in Cowper (7) or in any textbook on building construction published between say 1875 and 1920. The modern implications are seen in:

10. Spence, R.J.S. *Alternative Cements in India* Intermediate Technology Development Group, London, 1976

11. Bandyopadhyay, S., Kulkarni, S.G., Rajagopalan, B., 'An Alternative Cement for Mass Houses' *Lime*, Vol. 2. Khadi and Village Industries Commission, Bombay 1975

12. Spence, R.J.S., 'Lime and Surkhi Manufacture in India' *Lime and Alternative Cements* Intermediate Technology Publications Limited, London 1975

References for Chapter 2
Boynton (3) elaborates on the process of dissociation and on the matters which effect the reactivity of the limes produced. Cowper (7) uses reactivity as an indication of the nature of impure limes as suggested by the early nineteenth century French military engineer Vicat who carried out much of the early research on hydraulic limes. He published many books and papers, a late work was:

13. Vicat, L.J. *A Practical and Scientific Treatise on Calcareous Mortars and Cements, Artificial and Natural* Translated by Smith, J.T., London 1837

For information on thermal aspects, kiln aerodynamics and efficiency see the paper available from Intermediate Technology Industrial Services:

14. White, J.E. *Lime Manufacture* Intermediate Technology Industrial Services, Rugby 1981

References for Chapter 3
Raw materials are consindered by Boynton (3), Searle (1), Knibbs (2) and Bessey (6) as well as Stowell (4). Other sources of information are:

15. Boynton, R.S. and Gutschiek, K.A. *Building Lime: Its Properties, Uses and Manufacture* UNIDO Paper ID/WG.20/1 1968

16. Schwarzkopf, F. *Lime Burning Technology* Kennedy Van Saum Corporation Pennsylvania, USA. 1974

17. North, F.J. *Limestones: Their Origins, Distribution and Uses* London 1930

18. Dixey, F. *The Limestone Resources of Nyasaland* Bulletin No: 3, Geological Survey, Zomba, Nyasaland (now Malawi) 1927

19. Spence, R. *Appropriate Industrial Technology* UNIDO Monograph No. 12., Vienna 1980

20. Hill, N.R. *Factors in the Selection of a Site for a Factory for Building Material Production* Technical Report No: 15, UNDP/UNIDO Project INS/74/034. Bandung, Indonesia. December 1977

21. Hill, N.R. *Guide to the Geological Investigation of Rock*

Deposits for Building Material Production Technical Report No: 15, UNDP/UNIDO Project INS/74/073. Bandung, Indonesia. February 1977

22. Anon., *Economic Report No. 1*. Energy and Mineral Development Branch, Centre for Natural Resources, Energy and Transport, UN.NY 10017. May 1979

23. Government of Malawi *The Mines and Minerals Act, 1981* Malawi Gazette Supplement, Zomba, Malawi. 1st June, 1981

References for Chapter 4
Searle (1) gives a little advice on the use of wood for lime-burning and there is a passing comment in Boynton (3). Case studies are given in:

24. Sauni, J.T.M. and Sakula, J.H. *Oldonyo Sambu Pozzolime Industry: History, Operation and Development* Small Industries Development Organization. Dar es Salaam 1980

25. Ryan, W.H. ' "Coralite" for Cheaper Building in the Islands' *South Pacific Bulletin* July 1961

Other information can be found in:

26. Davey, N. *A History of Building Materials* London, 1961

27. Openshaw, K. *Woodfuel – a time for reassessment* Natural Resources Forum 3, United Nations, New York 1978

28. Earl, D.E. *Charcoal* Food and Agriculture Organization of the United Nations. Rome 1974

29. National Academy of Sciences *Firewood Crops* Washington D.C. 1980

30. *Public Works Department Handbook* Bombay 1948

31. Small Industries Development Organisation *Proposal for Afforestation Project: Oldonyo Sambu Ward* Dar es Salaam 1978

32. Sakula, J.H. *Lime and Pozzolana in Vanuatu* ITIS Rugby 1981

33. Booth, H. *Charcoal in the Energy Crisis of the Developing World* Food and Agriculture Organization of the United Nations. Rome 1979

34. Burley, J., Sandels, A. and Harker, S.P. 'Calorific Values of Wood and Bark' *Commonwealth Forestry Review, 60*, 3, 1981

35. International Labour Organisation *Appropriate Technology in Phillipine Forestry* Geneva 1977

General information on coals, oils and gases is available in standard reference books such as:

36. Kirke-Othmer *Encyclopaedia of Chemical Engineering* (1965 etc.)

The classification by Dryden was quoted from this source. The use of fluidized beds for burning coal and other fuels is described in:

37. National Coal Board *Fluidized Bed Combustion of Coal* London 1980

A brief note on the briquetting of waste materials was a received from an equipment manufacturer:

38. SPM Group Inc. *Energy and New Products From Waste Material* Englewood, Colorado, USA. 1983

References for Chapter 5
Searle (1) and Knibbs (2) describe many different types of kiln. The more modern high technology kilns are described by Boynton (3). There are few descriptions of the very simple kilns. Uncovered heaps are shown in the GRET study (5) and a covered heap is described by Wyatt Papworth in:

39. Gwilt, J. *An Encyclopaedia of Architecture* Book II, Chapter 11, Revised by Wyatt Papworth London 1888

The 'coralite' article in reference (25) describes a pit kiln and Hill has prepared another description of pit kilns in Central Sulawesi, Indonesia.

40. Hill, N.R. *Pit Kilns for Lime Production*

The rare dome kiln is fully described in an unpublished article by Stefan Cramer.

41. Cramer, S. *A Local Artisan at Work. Lime Burning on Maio Island, Cape Verde*

The simple, traditional, kilns are described in many old

encyclopaedias and in occasional specialist works such as:

42. Fourcroy de Ramecourt *Art du Chamfourmier* Paris 1766

They are also described in works of industrial archeology:

43. Grant, R. *Lime Kiln Construction* Resource Use Institute, Pitlochry, Scotland 1973

44. Aldsworth. F. *Archeology in West Sussex. Limeburning and the Amberley Chalk Pits* West Sussex County Council, Chichester 1979

Operational descriptions, in the context of scope for improvement, were given in an Indian study at the Central Building Research Institute, Rourkee:

45. Dave, N.G. and Masood, I. 'Studies on the existing lime kilns in India' *Lime: Manufacture and Uses* Lime Manufacturers' Association, Delhi 1972

and three examples are given in a short article:

46. Eeckhoudt, D. 'Documents – Fours a Chaux en Tunisie' *Environment Africain* No: 11–12 (Vol III, 3–4) Dakar 1979

The earlier improved kilns such as the ring kilns are described in Searle (1) and another ring kiln is described by Aldsworth (44). The recent improvements in small scale kilns are generally described in individual project reports. Designs for the two KVIC shaft kilns are published in:

47. Khadi and Village Industries Commission *Lime Kiln Designs* Bombay

Further information on developments from this design is given in Spence (10) and Sauni and Sakula (24). A very simple modern shaft kiln was described by:

48. Mason, S. *Shaft Lime Kiln* Papua New Guinea Department of Public Works, Building Research Station. Boroko 1974

The experimental batch kiln in Ghana was described by:

49. Ellis, C.I. 'Small Scale Lime Kiln Manufacture in Ghana' Paper 12 *Lime and Alternative Cements* Intermediate Technology Development Group, London 1975

The experimental batch kiln in Belize was explained in:

50. Halcrow Caribbean Limited *Lime Burning Investigations in Belize, Central America* 1977

The inclined chimney kiln in Honduras was based on a pottery kiln from the remarkable book:

51. Rhodes, D. *Kilns. Design, Construction and Operation* London 1970

The project was outlined in Vita News:

52. Anon. "Honduran Lime Kiln Test Fired" *Vita News* Vol. V, No. 3 Winter 1976

and described more fully in a conference paper:

53. Fox, T.H. Lime Production in Honduras. Case presented by VITA at the Lily Endowment Inc. – ITDG Conference on the Effective Use of Appropriate Technologies. Case Study No. 21 April 1977.

A later critique was included in a schools textbook:

54. Blum, J.E. 'Honduras: An Experimental Lime Kiln' *Appropriate Technology for Development – Case Histories*. Editors Evans, D.D., and Adler, N.L.

References for Chapter 6
Most of the references for advice on kiln design give examples rather than design guidance. Searle (1) is strong on examples. Other works relevant here are Bessey (6), Halcrow (50), Spence (10), Sauni and Sakula (24), KVIC (47).

An important detail is covered by:

55. Rao, B.L.K. 'Role of Refractories in Limekilns' *Lime* Vol 2. KVIC Bombay March 1975

But the most helpful advice is contained in:

56. Indian Standards IS 1849–1967 *Code of Practice for Design and Installation of Lime-kilns* Delhi 1967

which deals with small vertical mixed feed kilns only.

References for Chapter 7
Boynton (3) deals at length with the theory of hydration but his practical advice is restricted to high technology methods. Stowell gives good descriptions of the two major alternatives. Practical advice is given in:

57. Indian Standard IS 1635–1975 *Code of Practice for Field Slaking of Building Lime and Preparation of Putty* Delhi 1975

58. Dave, N.G., Mehrotra, S.P. and Khalid, M. 'Hydrated Lime: Production and Properties' *Lime: Manufacture and Uses* Lime Manufacturers Association of India Delhi 1972

Advice on the hydration of magnesian limes is given by Cowper (7) and:

59. Dave, N.G. and Masood, I. 'Investigations on Magnesian Limes' *Lime Manufacture and Uses* L.M.A. of India Delhi 1972

There are case studies in Sauni and Sakula (24) and:

60. Fewster, P.R. 'Lime Production at Swaneng Hill School, Botswana' Paper 11 in *Lime and Alternative Cements* ITDG 1975

There are references to small scale mechanical hydrators although neither of these given offer a full description:

61. Hill, N.R. *Mission Report: Technical Assistance to the Development of Building Material Production in the Southern District, Governemnt of Botswana.* UNIDO Vienna, 1982

62. Jain, S.K. 'A New Lime Hydrator' *Karamantha* India September 1978

References for Chapter 8
Guidance on small scale quarrying is given in:

63. Lester, D. *Quarrying and Rockbreaking* ITDG, London 1981

Lime-works operations on a very small scale are described by Fewster (60) and still on a village scale by Sauni and Sakula (24). A good, if very simple, book about kiln operation was produced by Claira:

64. *CLAIRA Looks at Lime-Kilns* The Chalk Lime and Allied

Industries Research Association, Welwyn, England (undated)

Periodicals
Research papers and reports on the lime industry have been reported in *Pit and Quarry* (US), *Rock Products* (US), *Zement-Kalk-Gyps* (W. Germany), *Tonindustrie-Zeitung und Keramische Rundschanl, Journal of Applied Chemistry* (London), *Indian Journal of Technology, Revue des Materiaux de Construction, Science Today* (Bombay), and *Appropriate Technology* (London).

12.2 Organizations

(a) Several countries now have appropriate technology organizations with some experience in lime-burning:

France: Groupe de Recherche et d'Echanges Technologiques (GRET), 34 rue Dumont d'Urville. 75116 Paris

Netherlands: Centre for Appropriate Technology, Sterinweg 1, (Kab., 6.33), Postbus 5048, 2600 Ga Delft (P.B.) Netherlands

West Germany: German Appropriate Technology Exchange (GATE), Postfach 5180, D–6236 Eschborn 1

United Kingdom: Intermediate Technology Development Group Limited, Myson House, Railway Terrace, Rugby CV21 3HT

USA: Volunteers in Technical Assistance (VITA), 3706 Rhode Island Avenue, Mt. Rainier, Maryland, USA 20822

(b) There are also some national and regional technical institutions:

Australia: CSIRO, Building Research Division, Melbourne

Ghana: Building and Road Research Institute, University of Science and Technology, University Post Office B0240, Kumasi

India: Khadi and Village Industries Commission, Irla Road, Vile Parle (West), Bombay

India: The Central Building Research Institute, Rourkee, UP 247672

Indonesia: Ceramic Research Institute, Jalan Jend. A. Yani 392, Bandung

Tanzania: National Housing and Building Research Unit, P.O. Box 1964, Dar es Salaam

Tanzania: Small Industries Development Organization, P.O. Box 2476, Dar es Salaam

UK: Building Research Establishment, Overseas Division, Garston, Watford Hertfordshire WD2 2JR

UK: Transport and Road Research Laboratory, Overseas Unit, Crowthorne, Berkshire RG11 6AU

(c) The United Nations Industrial Development Organization has taken an interest in small-scale lime-burning and has sponsored field studies on aspects of the subject. Refer to the Chemical Industries Branch, United Nations Industrial Development Organization, Vienna International Centre, P.O. Box 300, A–1400 Vienna

Appendix 1: Properties of the components of pure limes

Chemical name	Components of pure quicklimes		Components of pure hydrated limes	
	Calcium oxide	Magnesium oxide	Calcium hydroxide	Magnesium hydroxide
Chemical formula	CaO	MgO	Ca(OH)$_2$	Mg(OH)$_2$
Molecular weight	56.08	40.32	74.10	58.34
Specific gravity (approximate)	3.4	3.65	2.34	2.4

Appendix 2: Physical properties of limestone and limes

	Limestones		Quicklimes		Hydrated limes	
	High Ca	Dolomitic	High Ca	Dolomitic	High Ca	Dolomitic
True density (kg/m³)	2650 to 2750	2750 to 2900	3000 to 3400	3500 to 3600	2300 to 2400	2400 to 2900
Bulk density (kg/m³)	2000 to 2600	2050 to 2870	790 to 960	820 to 990	400 to 641	—
Angle of repose	—	—	50° – 55°	50° – 55°	70° wide range	70° wide range
Specific heat	0.22 mean	0.22 mean	0.22 mean	0.27 mean	around 0.3	around 0.32
Solubility (in water)	very low	very low	—	—	0.18% at 10°C	less than Ca (OH)$_2$

Limestones and their limes vary greatly and all of these figures are very approximate. Most of the information is drawn from *Boynton*, who describes many other properties.

Appendix 3: Time needed for dissociation

The relationship between lump sizes, temperature and time needed for dissociation was investigated by C.C. Furnass. He used an empirical equation to correlate his observed data and from this the following graph has been prepared. For example, curve A shows that a limestone sphere of diameter 25mm will dissociate into quicklime and CO_2 after about half an hour of heating at 1150°C.

The laboratory tests were made on limestone spheres suspended on fine tungsten wires. This method gives a 100% active surface which is a condition impossible to attain in any practical kiln and allowance for this must be made in any interpretation of the graph. But the graph does allow an estimate of the dissociation time for any given conditions of size and temperature.

Assume that the active surface for a reasonably well graded broken limestone is the area of the mean cube plus about 25% as a rough 'shape factor'. Compare this with the nearest size of sphere on the graph and read off the time against the working gas temperature.

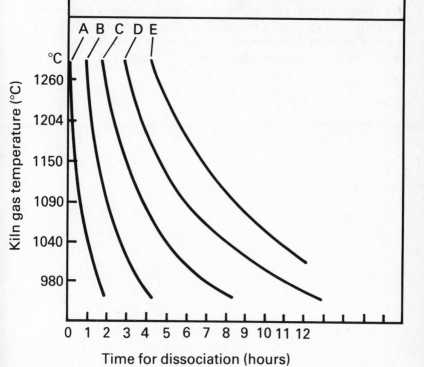

CURVE A 25mm diameter sphere,
surface area 1964mm^2
CURVE B 50mm diameter sphere,
surface area 7854mm^2
CURVE C 75mm diameter sphere,
surface area 17671mm^2
CURVE D 100mm diameter sphere,
surface area 31415mm^2
CURVE E 125mm diameter sphere,
surface area 49087mm^2

Appendix 4: Equilibrium pressure and temperature for dissociation of Calcium Carbonate

The release of carbon dioxide gas from the carbonate to form quicklime is a *reversible* reaction, so under some conditions the newly formed quicklime and the carbon dioxide may recombine. The equilibrium point of this reaction depends on the temperature and the partial pressure of the CO_2. The graph opposite, based on the work of Johnson and Mitchell, shows the equilibrium pressure for a range of temperatures. For lime-burning to proceed either the pressure must drop or the temperature must rise to a level above the equlibrium level. For example, at a temperature of 750°C the CO_2 partial pressure must be kept below about 75mm Hg.

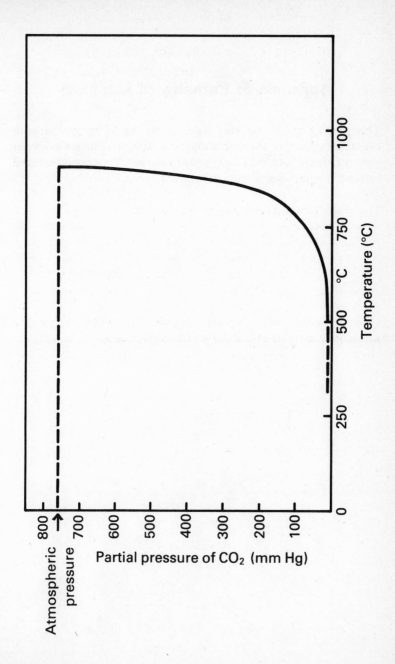

Appendix 5: Enthalpy of kiln gases

This graph provides an easy way of arriving at an approximate estimate of gas flow through a kiln (see App. 6). The enthalpy, or energy content, of the kiln gas shows how much energy is required to raise the gas to a given temperature.

The graph is based on a gas composition of:

CO_2	30%
H_2O	4%
air	15% (excess)
N_2	51%

However, even quite substantial variations from this composition will have very little effect on the enthalpy values.

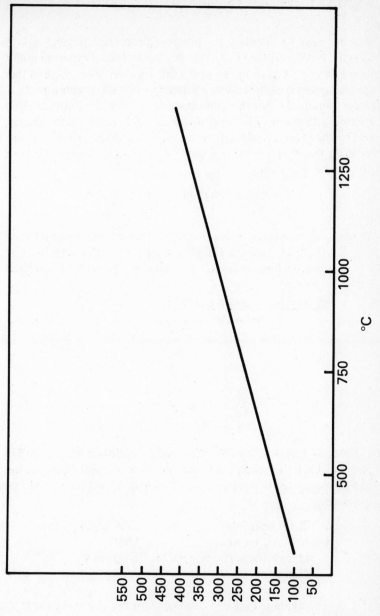

Appendix 6: Pressure drop through a kiln

It is not easy to calculate the pressure drop (Δp) through a kiln directly, but if a friction factor has been determined experimentally then Fanning's equation can give a fair forecast. The graph in this appendix was completed using Fanning's equation with a friction factor obtained from the laboratory work of C.C. Furnass. This factor is affected by the shape, roughness and packing of the charge and by the presence of the fuel in a mixed feed kiln. However, the plots on this graph provide a useful first estimate of the pressure drop to be expected.

$$\text{Fanning's equation is: } \Delta p = \frac{4f\ LV^2}{2gd}$$

Where: Δp = total pressure drop; f = friction factor (dimensionless); L = length (height) of kiln; V = gas velocity; g = gravitational constant; d = average diameter of kiln shaft (empty).

CURVE No.	SPACE VELOCITY	
	$m^3/min/m^2$	ft/min
1	30.45	100
2	27.33	90
3	24.32	80
4	21.30	70
5	18.18	60
6	15.23	50
7	12.28	40
8	9.14	30
9	6.09	20

Cold gas flow rate can be estimated from the enthalpy graph in Appendix 5, if the heat input to the kiln is known, and the maximum gas temperature. If the latter is not known, 1260°C is a good first estimate.

Example:
Total heat input   1364 kcal/kg lime
Kiln gas temp   1260°C
Gas enthalpy (from App. 5)   370 kcal/kg
Gas mass flow = $\dfrac{1364}{370}$ = 3.7 kg/kg lime

Using an average figure of 1.28kg/m³ for the gas density at 15°C, we get a cold gas flow rate of 3.7/1.28 = 2.89 m³ per kg lime made.

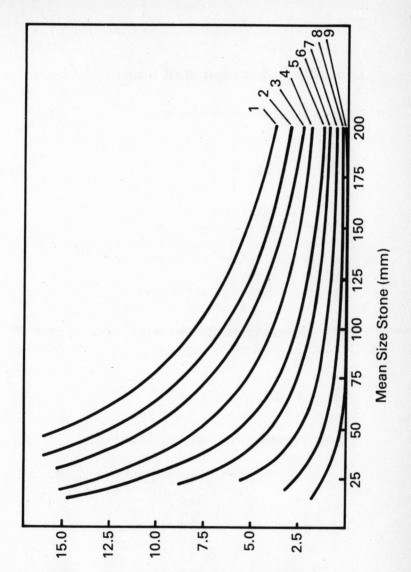

Appendix 7: Estimation of natural draught

The standard equation for chimneys is:

$$D = 1.215\,H \left(1 - \frac{w_g(T_a + 273)}{w_a(T_g + 273)}\right)$$

Where:

D = Draught (mm w.g.)
H = Height of kiln (m)
T_a = Ambient air temperature (°C)
T_g = Mean kiln gas temperature (°C)
w_a = Specific gravity of air at T_g
w_g = Specific gravity of kiln gases at T_g

The kiln gas mean temperature is likely to lie in the range of 500°C to 700°C, and at this temperature w_g very nearly equals w_a so:

$$D = 1.1215\,H \left(1 - \frac{(T_a + 273)}{(T_g + 273)}\right)$$

For quick reference, a few values are tabulated:

Height of kiln (m)	Draught (mm w.g.) for various T_g (°C)			
	550°	600°	650°	700°
5	3.9	4.1	4.2	4.3
6	4.7	4.9	5.0	5.1
7	5.5	5.7	5.8	6.0
8	6.3	6.5	6.7	6.8
9	7.1	7.3	7.5	7.7
10	7.9	8.1	8.4	8.5

Appendix 8: Theoretical heat requirement for lime-burning

The heat required for dissociation of calcium carbonate was explored by C.C. Furnass in the 1930s and a mean figure can be taken as 2817 kJ/kg (673 kcal/kg). The figure for pure dolomite would be around 6% lower.

The effect of small quantities of impurity may be ignored but, in general, any increase in impurities tends to *decrease* the heat required for dissociation.

The following table shows the heat balance for a very high technology German kiln quoted by Boynton:

	%	kJ/kg
1. Heat of dissociation	91.4	2875
2. Heat to dry the stone	1.5	47
3. Heat lost in dust and incombustibles	1.6	50
4. Sensible heat lost to discharged lime	1.1	35
5. Radiation from the kiln	1.3	41
6. Heat lost in exhaust gases	3.1	97
TOTAL	100	3145

The thermal efficiency in this exceptional example was 91.4%. Other examples of efficiency are 32.5% for a standard rotary kiln (Boynton), 46.9% for rotary kiln with heat recuperation, 55% for an oil fired shaft kiln in India and 39% for a large, though old, Solvay design mixed feed shaft kiln burning coke.

Appendix 9: A simplified Sayler chart showing various properties of coals

Several properties of coal are interrelated. It is possible to construct a chart showing:

% Carbon
% Hydrogen
% Volatile matter
Gross calorific value.

So that any two known values can be used to predict the others. the common names of various coals have also been shown on the chart included here. The normal commercial coals could be plotted within the two lines forming the band across the chart. The low ranking coals are at the top and the high ranking coals (perversely) at the bottom.

In section 4.4 when producer gas was considered, certain preferred characteristics were suggested. These are shown by the dotted area within the bituminous coals range.

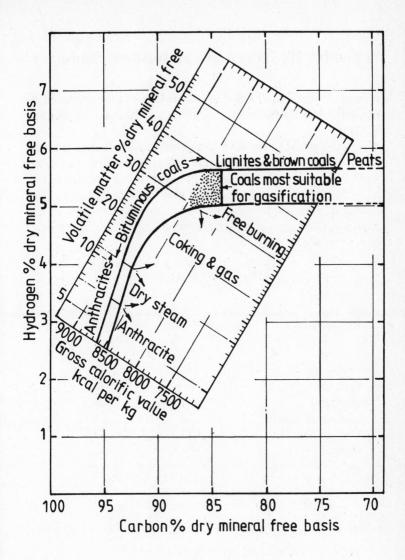

Appendix 10: Notes on combustion chemistry

The way in which fuels burn in a lime-kiln can be predicted by considering only the two main constituents of the fuel — carbon and hydrogen.

The simplest way to handle kiln combustion calculations is on the basis of weights and in round terms *3.7kg of combustion gases are needed for each 1kg of lime*. Better quality lime may be made using slightly more gases and cheaper lime may be made by cutting back the weight of gases to perhaps 3.6kg. For large kilns and for sensitive applications, such as lime for chemical uses, more exact calculations will be needed.

Molecular Weights

Element	Symbol	Molecular Weight
Carbon	C	12
Hydrogen	H	2
Oxygen	O	32
Nitrogen	N	28
Carbon monoxide	CO	28
Carbon dioxide	CO_2	44
Water	H_2O	18

Carbon burning partially to carbon monoxide

Carbon + Oxygen + (Nitrogen) = Carbon Monoxide + (Nitrogen)

$$C + 1/2 \, O_2 + (1.88 \, N_2) = CO + (1.88 \, N_2)$$

$$12 + 16 + (52.64) = 28 + (52.64)$$
$$1 + 1.3 + (4.4) = 2.3 + (4.4)$$

So each 1kg of carbon burns in air to yield 2.3kg of carbon monoxide and leaves 4.4kg of nitrogen. The nitrogen takes no part in the reaction but is, of course, present in the air.

The reaction is the partial combustion of carbon to carbon monoxide which occurs in the 'starved air' conditions in a gas producer.

Carbon monoxide burning to carbon dioxide

$$CO + 1/2\ O_2 + (1.88\ N_2) = CO_2 + (1.88\ N^2)$$

$$28 + 16 + (52.6) = 44 + (52.6)$$

$$1 + 0.57 + (1.88) = 1.57 + (1.88)$$

So each 1kg of carbon monoxide burning to carbon dioxide in air produces 1.57kg of carbon dioxide and leaves 1.88kg of nitrogen.

Carbon burning directly to carbon dioxide

$$C + O_2 + (3.67\ N_2) = CO_2 + (3.67\ N_2)$$

$$12 + 32 + (105.3) = 44 + (105.3)$$

$$1 + 2.66 + (8.77) = 3.66 + (8.77)$$

This reaction, the complete combustion of carbon in one stage, occurs in all limekilns other than those fired with producer gas.

Hydrogen burning to water

$$H_2 + 1/2\ O_2 + (1.88\ N_2) = H_2O + (1.88\ N_2)$$

$$2 + 16 + (52.6) = 18 + (52.6)$$

$$1 + 8 + (26.3) = 9 + (26.3)$$

So each 1kg of hydrogen burning in air to water produces 9kg of water (as vapour in the kiln) and leaves 26.33kg of nitrogen.

Example

Consider the burning of natural gas as a fuel for a lime-kiln. The analysis of samples showed this composition:

$$86.5\%$$
$$13.2\%$$
$$0.3\%$$

The trace elements may be ignored and each 1kg of natural gas contains:

Carbon   0.865kg
Hydrogen 0.132kg

From the notes above it can be seen that

0.865kg carbon yields
  $(0.865 \times 3.66) \, CO_2 + (0.865 \times 8.77) \, N_2$
  $= 3.27kg \, CO_2 + 7.58kg \, N_2$

and

0.132kg hydrogen yields
  $(0.132 \times 9) \, H_2O + (0.132 \times 26.33) \, N_2$
  $= 1.19kg \, H_2O + 3.48kg \, N_2$

So 1kg of the natural gas yielded
  $(3.27 + 7.58 + 1.19 + 3.48)$
  $= 15.5kg$ of kiln combustion products.

The following table summarizes the products obtained by burning carbon, carbon monoxide, hydrogen, methane and sulphur both on a weight basis and on a volumetric basis.

Air required for combustion yielded on a weight basis

Constituent	Symbol	Theoretical air requirement kg per kg	Products				
			CO_2	H_2O	N_2	CO	SO_2
Carbon (to CO_2)	C	11.49	3.67	—	8.82	—	—
Carbon (to CO)	C	5.75	—	—	4.42	2.33	—
Carbon monoxide	CO	2.46	1.57	—	1.89	—	—
Hydrogen	H_2	34.48	—	9.00	26.48	—	—
Methane†	CH_4	17.24	2.75	2.25	13.24	—	—
Sulphur	S	4.31	—	—	3.31	—	2.00

Air required for combustion and products yielded on a volume basis

Constituent	Symbol	Theoretical air requirement m^3 per m^3	Products					
			CO_2	H_2O	N_2	CO	SO_2
Carbon monoxide	CO	2.38	1.00	—	1.88	—	—
Hydrogen	H_2	2.38	—	1.00	1.88	—	—
Methane†	CH_4	9.52	1.00	2.00	7.52	—	—

In the above tables, all data is reduced to two places of decimals: This is quite accurate enough for all practical lime-kiln calculations.
† Natural gas may be treated as methane.

Appendix 11: Physical and chemical properties of typical samples of five principal grades of industrial fuel oil

		Light distillates		Heavy industrial fuel oils		
		Kerosene	Gas oil	Light fuel oil	Medium fuel oil	Heavy fuel oil
Relative Density at 15°C	litres per tonne	0.790 1254	0.835 1175	0.930 1075	0.940 1043	0.960 1029
Kinematic viscosity at 15°C 3.78°C 82.2°C	centistokes centistokes centistokes	2.0	3.0	12.5	30.0	70.0
Minimum storage temperature	°C			10	25	35
Minimum temperature at tank outflow	°C			10	25	45
Gross calorific value Nett calorific value	kJ/kg kJ/kg	46,500 43,900	45,500 43,000	43,500 41,000	43,000 40,700	42,800 40,700
Composition by weight — Sulphur (S) Carbon (C) Hydrogen (H) Water, ash, etc.	% weight % weight % weight % weight	0.1–0.2 86.3 13.6 —	0.6–0.9 86.3 12.8 —	1.4–2.3 86.2 12.4 —	2.1–2.4 86.1 11.8 Trace	2.5 87.3 9.5 1.0

Appendix 12: Properties of gaseous fuels

		Liquid phase		Vapour phase		Natural gas	Producer gas from coal	Producer gas from fuel oil
		Propane	Butane	Propane	Butane			
Density	kg/m^3	510	575	1.86	2.46	0.67	1.03	—
	m^3/tonne	1.96	1.74	537	406	1480	970	—
Calorific value gross	kJ/kg	6070	6790	50350	53700	53000	5780	4810
nett	kJ/kg	5620	6270	46400	45800	47850	5420	4560
Carbon C	% weight	81.72	82.66	81.72	82.66	76.1	20.0	6.81
Hydrogen H		18.28	17.34	18.28	17.34	23.0	1.20	0.49
Nitrogen and incombustables		trace	trace	trace	trace	0.3–0.9	78.8	84.49
Methane CH$_4$		—	—	—	—	86.5	—	1.97
Higher alkanes		—	—	—	—	13.2	—	—
Carbon monoxide CO	% volume	—	—	—	—	—	28.0	12.0
Carbon dioxide CO$_2$		—	—	—	—	—	4.5	—
Oxygen O$_2$		—	—	—	—	—	0.6	—
Hydrogen H$_2$		—	—	—	—	—	14.0	10.0
Nitrogen N$_2$		—	—	—	—	—	51.0	77.1
Methane CH$_4$		—	—	—	—	—	3.0	3.0

Appendix 13: Height of kiln for various stone sizes

The main plot (page 184) shows pressure drop per metre of bed depth for different stone sizes at varying gas flow rates.

Assuming a usual working rate of 30Std m^3/min per m^2, the data for the smaller plot (page 185) can be read off, and if a total acceptable kiln pressure drop of 200mm w.g. is taken (much too high for small kilns) the total bed depth can be calculated. This gives the smaller plot, which extrapolates to zero as expected.

The line 'A' was added as a convenience since it shows how draught improves heat transfer and therefore output; but draught means power which is not usually available for small kilns, and mixed feed kilns do not respond well to high draught. Even a 4 tonne/hr Solvay kiln with fully automatic charge/discharge gear will only stand 50–60mm w.g. draught.

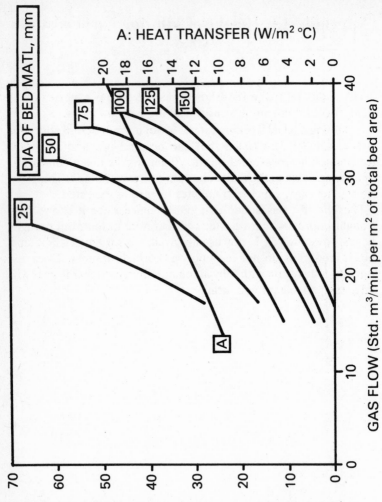

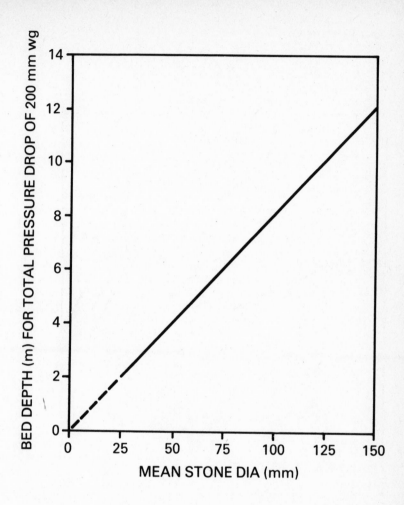